Electromagnetic Compatibility

Yang Zhao · Wei Yan · Jun Sun · Mengxia Zhou ·
Zhaojuan Meng

Electromagnetic Compatibility

Principles and Applications

Science Press
Beijing

Springer

Yang Zhao
Nanjing Normal University
Nanjing, Jiangsu, China

Wei Yan
Nanjing Normal University
Nanjing, Jiangsu, China

Jun Sun
Experiment and Verification Center
State Grid Electric Power Research Institute
Nanjing, Jiangsu, China

Mengxia Zhou
Nanjing Normal University
Nanjing, Jiangsu, China

Zhaojuan Meng
Nanjing Normal University
Nanjing, Jiangsu, China

ISBN 978-981-16-6454-0   ISBN 978-981-16-6452-6   (eBook)
https://doi.org/10.1007/978-981-16-6452-6

Jointly published with Science Press
The print edition is not for sale in China (Mainland). Customers from China (Mainland) please order the print book from: Science Press.

This Springer imprint is published by the registered company Springer Nature Singapore Pte Ltd.
The registered company address is: 152 Beach Road, #21-01/04 Gateway East, Singapore 189721, Singapore

Preface

Electromagnetic compatibility (EMC) is a comprehensive interdisciplinary subject that infiltrates and combines with many other subjects. It is an interdisciplinary subject of natural science and engineering with broad theoretical foundation and comprehensive engineering practice, which is also a requirement for engineers.

With the rapid development of electrical and electronic technology, EMC testing methods have gained more and more attention from electronic, electrical engineers, and engineering technicians. In order to ensure the stability and reliability of the equipment and meet the requirements, designers need to design and test the EMC system precisely.

Compared with the countries with advanced science and technology in the world, China started late and falls a large gap behind. At present, in order to make domestic electrical equipment and electronic products sell in the domestic market and enter the international market, the products must pass and comply with the mandatory testing and inspection standards for electromagnetic compatibility. Therefore, it is necessary for college students and the majority of electrical and electronic engineers and technicians to accept theoretical and technical education on EMC. However, for a long time, the domestic technical reference materials on EMC have not been comprehensive. For college students and designers, there are few EMC textbooks and reference materials with characteristics and close integration of theory and engineering practice. This book is written for this purpose and is mainly used as a textbook or reference material for EMC training and learning for graduate students in electrical engineering majors and electrical engineers.

This book is composed of several authors who combine the latest research results and professional theory to solve practical engineering problems. It is mainly reflected in the following aspects:

(1) The engineering practicability and feasibility of EMC mechanism analysis method are reflected in a large number of actual cases.
(2) The test method combines the latest research results of the authors, which is reliable in theory and feasible in practice.

Thanks to the postgraduates such as Bai Wanning, Zhu Zhibo, Wang Yongan, Huang Junshuo, Wu Xianqiang, Hou Shiliang, Chen Jian, Fang Juhao, Liu Qiangqiang, Huang Chao, and Ji Yanxing who have done a lot of editing and proofreading work. Thanks to the authors of the references in the book.

I sincerely thank Nanjing Normal University for providing a solid foundation for my scientific research path. Thanks for the support and assistance provided by the Jiangsu Society for Electrical Engineering and the State Grid Electric Power Research Institute of NARI Group. Furthermore, I thank my graduate students, Zhu Zhibo, Bai Wanning, and my secretary Gu Xiaoyi, for their hard work and dedication during the editing of this book, which promoted the rigor and richness of the book.

Due to the rush of time and the limited level of authors, it is unavoidable that there are improprieties or errors in the book, and readers are invited to criticize and correct them.

Nanjing, China   Yang Zhao
November 2020

Contents

Chapter 1
Introduction of Electromagnetic Compatibility

1.1 Electromagnetic Compatibility History and Basic Concepts

During World War II, the use of electronic equipment, especially radio transceivers, navigation equipments, and radars, led to an increase in the number of interference between radio transceivers and navigation equipments on aircraft. By redistributing the frequency of the transmission over a non-congested spectrum, or by keeping the cable away from the noise source to prevent cables from receiving interference, the interference problem can usually be resolved easily. At that time, the density of the electronics (mainly the electronic vacuum tube) was far less than today's. Therefore, it was easy to implement the modification of the interference on a case-by-case basis to solve the electromagnetic interference (EMI) problem. However, with the invention of high-density electronic components, such as field effect transistors invented in 1950s, integrated circuits (IC) invented in 1960s, and microprocessor chips invented in 1970s, the interference problem became increasingly serious. As the demands for voice and data transmission increase, the spectrum becomes more and more crowded, which requires a reasonable planning of the spectrum usage [1].

Due to the increasing number of digital systems that interfere with wired and wireless communications, the US Federal Communications Commission (FCC) issued a regulation in 1979, requiring electromagnetic interference from all "digital devices" to be lower than a certain limit. The purpose of the regulation was to limit "electromagnetic pollution," to prevent, or at least reduce the number of EMI cases. Because the digital devices sold in the USA had to meet the FCC mandatory limits, there was a strong interest in EMC disciplines among electronics manufacturers with products ranging from digital computers to electronic typewriters.

Many European countries had already enforced similar requirements for digital devices before the FCC promulgated the specifications. In 1933, the International Electrotechnical Commission (IEC) recommended the International Special Committee on Radio Interference (CISPR) to deal with emerging EMI issues at a meeting in Paris [1, 2]. The committee published a document detailing the measurement

© Science Press 2021
Y. Zhao et al., *Electromagnetic Compatibility*,
https://doi.org/10.1007/978-981-16-6452-6_1

equipment used to determine potential EMI problems. CISPR reconvened in London in 1946 after the end of World War II. Subsequent meetings published various technical publications, discussed measurement techniques, and suggested the limits. Some European countries have adopted the limits of various versions recommended by CISPR. The FCC specification was the first one for digital systems in the USA. The limits were based on CISPR recommendations and were subject to the change in the US environment. To prevent "field problems" associated with EMI, most electronics manufacturers had set internal limits and standards for their products in the USA. However, FCC specifications made such a voluntary act into a legal compliance program requirement.

These specifications have made EMC a key factor in market access for electronics. If a product does not meet the specifications in a certain country, it may not be sold in that country [2, 3]. In other words, products that are fully functionally implemented cannot be purchased as long as they do not meet the specifications.

Work on electromagnetic compatibility started late in China and developed gradually from 1970s. Some national standards and national military standards have been promulgated such as EMC design requirements and test methods, but the specific design specifications are still lacking. The work of electromagnetic compatibility penetrates into every electrical and electronic system and equipment. The electromagnetic compatibility problems can only be solved by management and coordination of overall design.

1.2   Electromagnetic Compatibility Standards and Measurement

Most electrical and electronic equipments, circuits, and systems emit electromagnetic energy, either intentionally or unintentionally. This emission can constitute electromagnetic interference. At the same time, many modern electronic devices, circuits, and equipments are capable of responding to or being affected by such electromagnetic interference. This problem has become more serious in modern semiconductor devices and VLSI circuits, which are prone to fail or even be completely damaged under electromagnetic interference because of their relatively low susceptibility thresholds for electromagnetic interference. Problems associated with electromagnetic emissions (constituting electromagnetic interference) and equipment, subsystems, and devices against electromagnetic interference (electromagnetic compatibility) are common in the production and transmission of wireless broadcast, communications, control, information technology products, instruments, computers, and electrical energy.

As a practical measure to ensure electromagnetic compatibility, the design and performance standards of various devices continue to evolve, and various organizations have been constantly releasing various relevant standards. The goal of these standards is to establish reasonable limits for electromagnetic emission levels and immunity limits for different devices. Electromagnetic interference or electromagnetic compatibility often involves weak signals or low interference levels, and the test procedure requires accurate measurements at very low power. In addition, different test procedures or different test instruments have different test results, although the differences may be small. Therefore, it is necessary to define the test process and test instruments carefully. Correspondingly, the standard also specifies test procedures and instruments for measuring electromagnetic (interference) emissions and susceptibility. The same instrument may show significant differences in testing at different test sites. In order to avoid difficulties in this area, sufficient attention should be paid in this respect.

1.2.1   FCC Standard

The Federal Communications Commission (FCC) is responsible for promoting and ensuring the effective implementation of various regulations in the USA involving radio broadcasting and communications facilities. The FCC also shoulders the standardization of electromagnetic emission control for various electrical and electronic equipment. It was enacted in the telecommunications regulations. The limits for the electromagnetic emissions (unintentional and intentional radiation) of radio frequency devices and equipment are specified.

1.2.2   CISPR Standard

Since 1930s, the European-based International Special Committee on Radio Interference (CISPR) has been actively engaged in the development of international standards for EMI/EMC which have been published by the International Electrotechnical Commission (IEC). The achievements of CISPR/IEC are international, involving not only EU countries but also other non-EU countries such as Australia, Canada, India, Japan, South Korea, and the USA. The IEC/CISPR documents on EMI/EMC are given in Table 1.1.

The methods of test and evaluation are the same as the ANSI/IEEE standards, and the IEC/CISPR documents and the standards are recommendations only. It allows participating countries and other countries to decide which part of the recommendations to implement and how to implement them in their countries. The description of the test platform and test procedures in the following sections is generally consistent with the tests in the corresponding IEC/CISPR standards. It is necessary to refer to the corresponding standards and follow the listed details and procedures strictly.

Table 1.1   IEC/CISPR standards for EMI/EMC

Theme	Standards
General	CISPR7B, CISPR8B, CISPR10
Measurement process and instrument use	CISPR16, CISPR17, CISPR19, CISPR20, CISPR8B, 8C, CISPR11, CISPR12, CISPR13, CISPR14, CISPR15, CISPR18-1, 2, 3, CISPR20
Performance limit	CISPR9, CISPR11, CISPR12, CISPR13, CISPR14, CISPR15, CISPR18-3, CISPR21, CISPR22

1.2.3 GB Standard

China also attaches great importance to the formulation and establishment of EMI protection and compatibility standards, because standardization is an important part of scientific management, and a technical basis for organizing modern production and promoting technological progress and technical exchanges with developed countries [3].

In China, the first EMC standard was issued by the former Ministry of Machinery Industry in 1966, that B, the Ministry of Standards JB 854-1966 "Marine Electrical Equipment Industry Radio Interference Terminal Voltage Measurement Method and Allowable Value." In the late 1970s, the Radio Interference Standardization Working Group was established under the auspices of the former National Bureau of Standards [4]. On October 31, 1983, the first EMC national standard GB 3907-1983 "Basic Measurement Method for Industrial Radio Interference" was promulgated. Subsequently, more than 30 national standards were issued such as GB 4343-1984 "Measurement Methods and Allowable Values of Radio Interference Characteristics of Power Tools, Household Appliances and Similar Appliances", GB 4365-1984 "Noun Terms of Radio Interference" and GB 4859-1984 "Basic Measurement of Anti-interference Characteristics of Electrical Equipment" [4–6]. These standards are relevant to IEC/CISPR standards, such as IEC/TC77 or IEC/TC65. In 1986, the National Technical Committee for Radio Interference Standardization, led by the State Bureau of Technical Supervision, was formally established and affiliated with the Shanghai Institute of Electrical Apparatus. The institute is responsible for the promotion and implementation of EMC standards [5–8].

Later, according to the needs of domestic work, the sub-technical committees corresponding to the IEC/CISPR/A.B.C.D.E.F.G were established, and the S-segment was specially established. There are eight official academic groups nowadays.

1.3   Electromagnetic Compatibility Terminology

Electromagnetic compatibility refers to the ability of a device or system to operate as required in an electromagnetic environment and to produce no electromagnetic noise to other surrounding equipment or systems. As a result, EMC covers two aspects: On

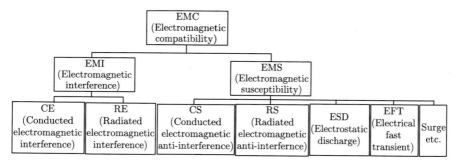

Fig. 1.1 EMC concept

the one hand, when the equipment performs, the electromagnetic interference to any other equipment or equipment around it must be kept below a certain limit [7–9]. On the other hand, the equipment has a certain degree of susceptibility (electromagnetic susceptibility). In other words, it can resist electromagnetic interference generated by any other surrounding electronic equipment to a certain extent.

EMC (electromagnetic compatibility) includes two parts: electromagnetic interference (EMI) and electromagnetic susceptibility (EMS), as shown in Fig. 1.1. EMI refers to the electromagnetic noise generated by equipment or a system during normal operation that affects other equipment or systems. EMS refers to the ability of equipment or a system to be susceptible to electromagnetic noise in the surrounding electromagnetic environment during normal operation.

Depending on the path of propagation, electromagnetic interference consists of two parts: conducted electromagnetic interference (conducted EMI) and radiated electromagnetic interference (radiated EMI). Further, the conducted EMI noise can be divided into common mode noise (existing between the live and ground lines, between the neutral line and the ground line) and differential mode noise (existing between the live and neutral lines). Radiated EMI noise propagates in the form of electromagnetic waves in space. It can also be roughly divided into common mode radiated noise and differential mode radiated noise. At the same time, the three elements of electromagnetic compatibility refer to the interference source, the coupling path (or transmission path), and the receptor [10]. As shown in Figs. 1.1 and 1.2, the interference source refers to the component and equipment that emits electromagnetic interference noise. The transmission path (or coupling path) refers to the medium that connects electromagnetic interference energy to sensitive equipment, including cables, space, etc. A sensitive device is a device that is affected by electromagnetic interference in an electromagnetic environment. The three elements of electromagnetic compatibility are the three conditions that must be possessed simultaneously for all electromagnetic interference to be generated.

According to different test methods, electromagnetic susceptibility consists of conducted interference test in radio frequency fields, radiated interference test in radio frequency fields, electrostatic discharge test, electrical fast transient test, surge (impact) test, etc.

Fig. 1.2 Three elements of
electromagnetic interference

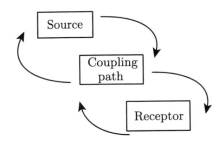

References

1. Paul CR. Introduction to electromagnetic compatibility. Hoboken: Wiley; 1992.
2. Henry WO. Noise reduction techniques in electronic systems. 2nd ed. Hoboken: Wiley; 1988.
3. Chatterton PA, Houlden MA. EMC: electromagnetic theory to practical design. Hoboken: Wiley; 1991.
4. Goedbloed JJ. Electromagnetic compatibility. J Am Soc Naval Eng. 1961;73(4):763–6.
5. Williams T. EMC for product designers. 3rd ed. Oxford: Elsevier; 2007.
6. Kodali VP. Engineering electromagnetic compatibility. Piscataway, NJ: IEEE Press; 1996.
7. Marshman C. The guide to the EMC directive 89/336/EEC. 2nd ed. London: EPA Press; 1993.
8. Morgan D. A handbook for EMC testing and measurement. London: IET Digital Library; 1994.
9. Liu PC, Qiu Y. Electromagnetic compatibility principle and technique. Beijing: Higher Education Press; 1993. (in Chinese).
10. Lin GR. Electromagnetic interference and suppression. Beijing: Electronic Industry Press; 2003. (in Chinese).

Chapter 2
Conducted EMI Noise Generation Mechanism, Measurement and Diagnosis

2.1 Generation Mechanism and Analysis of Conducted EMI Noise

2.1.1 Common Mode and Differential Mode Definition

Conducted electromagnetic interference noise can be divided into common mode noise (CM) and differential mode noise (DM) according to different generation mechanisms. Common mode noise refers to the noise propagating between the live (L) and ground (G) lines, and between neutral (N) and ground (G) lines. Differential mode noise refers to the noise propagating between the live (L) and neutral (N) lines.

1. Differential Mode Conducted EMI Noise

The differential mode signal is also called normal mode, series mode, line inductance, or symmetrical signal. In the two-wire cable transmission loop, each line-to-ground voltage is represented by symbols U_1 and U_2. The differential mode signal component is U_{DEF}. Pure differential mode signal is: $U_1 = -U_2$; the values are equal, and the phase difference is 180°. $U_{DM} = U_1 - U_2$, because U_1 and U_2 are symmetrical to the ground no current flows through the ground.

All differential mode currents (I_{DEF}) flow through the load. Differential mode interference invades two signal lines, the direction is consistent with the direction of the signal current, one of which is generated by the signal source, and the other is generated by electromagnetic induction during transmission, which is with the signal and is in phase. Interference is generally more difficult to suppress.

2. Common Mode Conducted EMI Noise

The common mode signal is also called the ground sense signal or the asymmetric signal, and the common mode signal component is U_{CM}, pure common mode signal is: $U_{CM} = U_1 = U_2$. The values are equal, and the phase difference is 0. When the interference signal invades between the line and the ground, the interference current flows through one-half of each line, and the ground is a common circuit. In principle,

© Science Press 2021
Y. Zhao et al., *Electromagnetic Compatibility*,
https://doi.org/10.1007/978-981-16-6452-6_2

this interference is easy to eliminate. Due to the unbalanced line impedance, the common mode signal interference is converted into crosstalk interference which is difficult to eliminate in the actual circuit.

For a pair of signal line a and b, the differential mode interference is equivalent to adding an interference voltage between a and b. The common mode interference is equivalent to adding an interference voltage between A and ground, B and ground, respectively; as usual, the differential signal is transmitted by twisted pair to eliminate common mode noise. The principle is very simple. The two wires are screwed together, and the common mode interference voltage is very close. $U_a - U_b$ remains the same value, which is the ideal situation. For example, the RS422/485 bus is a specific application that uses differential transmission signals.

2.1.2   Conducted EMI Noise Generation Mechanism

In practical applications, the influence of various environmental noises can be regarded as common mode interference, but if the noises of the two lines are not the same, the voltage between the two lines becomes different, which makes the common mode noise converted into differential mode noise. Differential mode signals are not necessarily relative to ground. If a line is grounded, their difference is relative to the ground [1]. This is the single-ended input of the differential circuit described in the analog circuit.

1. Common Mode Emission

The main contributor to common mode emissions is the parasitic capacitance from the primary side to ground. The three parts contributing to this capacitance are shown in Fig. 2.1, which is the capacitance of the switch crystal to the heat sink, the inter-winding capacitance of the transformer, and the primary side stray wiring capacitance [1, 2].

The single largest contribution is usually from the capacitance of the transistor to the heat sink. This capacitor can be reduced by: ① using an insulated thermal gasket containing a Faraday shield between the transistor heat sinks; ② using a thick ceramic ring (such as yttrium oxide); ③ a heat sink not grounded. The Faraday shielded insulated thermal washer consists of a copper shield between two thin layers of insulating material. In order to be effective, the copper shield must be connected to the source of the switching transistor. For bipolar switching transistors, the copper shield should be connected to the emitter. If the heat sink is electrically floating point grounded, it must be protected from contact with the heat sink for safety. In the event of an accident, the insulating thermal gasket between the switching transistor and the heat sink fails with a potential of the AC line, posing a risk of electrical breakdown.

The second contribution to this parasitic capacitance comes from the capacitance between the windings of the transformer. Because the designer wants a small transformer, the primary and secondary windings are placed closely together, which maximizes the inter-winding capacitance. A further transformer that separates the

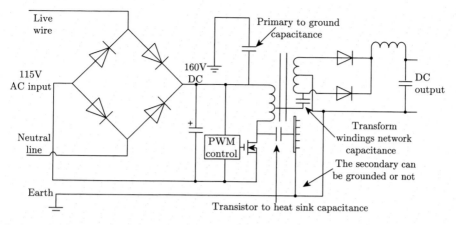

Fig. 2.1 Switching power supply with grounded parasitic capacitance

coils or a transformer that contains a Faraday shield reduces this capacitance. A disadvantage of transformers with Faraday shields is that they increase cost and may increase the size. Careful component placement, including careful routing and (or) printed circuit board (PCB) design, minimizes the third contribution, primary wiring capacitance.

Add LISN to the circuit of Fig. 2.1, and show the common mode conducted transmission path to get the circuit shown in Fig. 2.2. Pay attention to the common mode current, the LISN impedance looks like 25 Ω, that is, the parallel connection of the two 50 Ω resistors. The circuit of Fig. 2.2 can be greatly simplified by representing the switching transistor as a square wave voltage generator having a peak amplitude equal to the DC voltage across the filter capacitor C_F. The simplified common mode equivalent circuit of SMPS (switching mode power supply) is shown in Fig. 2.3.

From Fig. 2.3, it shows that the power supply has a high source impedance equal to the size of the capacitive reactance C_p. Therefore, the common mode current and the LISN voltage are mainly determined by the magnitude of this parasitic capacitance [2, 3]. Typical values range from about 50–500 pF. From the circuit in Fig. 2.3, the amplitude of the common mode voltage U_{CM} across the LISN resistor can be determined as:

$$U_{CM} = 50\pi f C_p U\left(f\right) \tag{2.1}$$

where $U(f)$ is the amplitude of voltage source at frequency U_p.

Since the voltage source is a square wave, the Fourier spectrum can be used to determine the harmonic components of the voltage. For a square wave, the envelope of the Fourier spectrum drops at a rate of 20 dB/dec to a frequency $1/\pi t_r$, where t_r is the rise/fall time of the switching transistor, beyond which it drops at a rate of 40 dB/dec.

The frequency term in Eq. (2.1) indicates an increase in 20 dB/dec. As a result of combining Eq. (2.1) and Fourier spectrum $U(f)$, the common mode conducted

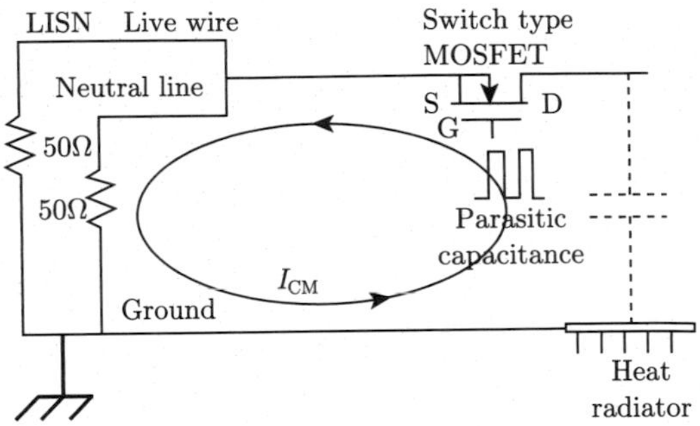

Fig. 2.2 Common mode equivalent circuit of switching power supply

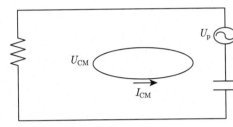

Fig. 2.3 Simplified common mode equivalent circuit of switching power supply

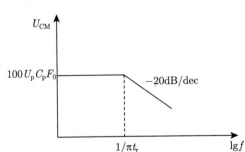

Fig. 2.4 Envelope of common mode conducted emissions varies with different frequencies

emission voltage U_{CM} is flat before the frequency $1/\pi t_r$, and the frequency decreases at the rate of 20 dB/dec, as shown in Fig. 2.4. The curves of Fig. 2.4 are limited to conducted emissions; however, the actual emissions are only present at the harmonics of the fundamental frequency F_0.

Since the emission is flat before the frequency $1/\pi t_r$ and then decreases at the rate of 20 dB/dec, the emission of a certain point (such as the fundamental frequency) is necessary to draw the complete envelope. The amplitude of the fundamental frequency is $0.64U_p$. Based on the Fourier series, substituting f into the fundamental frequency F_0, and substituting $U(f)$ into $0.64U_p$, there is the expression of the common mode conducted emission amplitude at the following fundamental frequency [3]:

$$U_{CM} = 100U_p F_0 C_p \tag{2.2}$$

It reveals that for the rise time of 100 ns, the turning point of the curve in Fig. 2.4 is at 3.18 MHz. Once the fundamental frequency of the switching power supply is selected, the only parameter that can be controlled by the designer is the parasitic capacitance C_p.

From Fig. 2.4, it is observed that slowing down the rise time of the switching transistor has an adverse effect on increasing the power loss, but does not reduce the maximum amplitude of the common mode conducted emission. The only effect of slowing down the rise time is to move the turning point to a lower frequency. This reduces high-frequency emission but not the maximum transmission of the low frequency.

2. Differential Mode Emission

When the power supply is working normally, the switching transistor drives a current along a loop at the switching frequency. This ring is composed of a switching transistor, a transformer, and a filter capacitor C_F. If the switching current flows through this loop within the power supply, no differential mode emissions can occur.

However, the main function of capacitor C_F is to filter the full-wave rectified AC line voltage. Therefore, the filter capacitor is a large-capacitance, high-voltage capacitor (usually with capacitance 1000 μF, rated voltage 250 V or higher) and is far from an ideal capacitor. There is usually a significant equivalent series inductance (ESL)L_F and an equivalent series resistance (ESL)L_F. As a result of the impedance, not all of the switching current flows through capacitor C_F. As shown in Fig. 2.5, a shunt appears at the capacitor, some of the switching current flows through the capacitor, and the rest flows through the full-wave bridge rectifier to the power line. The switching current flowing to the power line is the differential mode noise current flowing through the LISN. Note that for differential mode currents, LISN looks like 100 Ω (series of two 50 Ω resistors) [4]. The circuit of Fig. 2.5 can be simplified by replacing the switching transistor with a current generator I_p, and removing the bridge rectifier. The simplified differential mode equivalent circuit showing only the differential mode conducted emission current path is shown in Fig. 2.6.

From the circuit of Fig. 2.6, there is low differential mode source impedance due to the large capacitance of the input ripple filter capacitor C_F. Therefore, the differ-

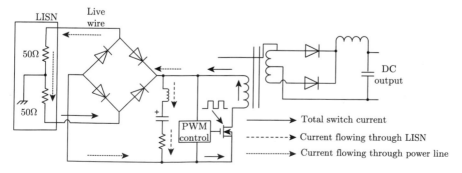

Fig. 2.5 Switching power supply with differential mode current path. *Note C_F has a shunt*

Fig. 2.6 Simplified differential mode equivalent circuit of switching power supply

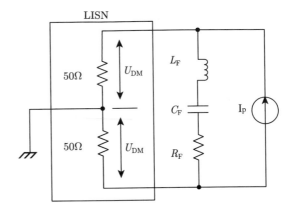

ential mode current and the LISN voltage are mainly determined by the installation of the parasitic parameters (L_F and R_F) and the filter capacitor C_F. Unreasonable installation increases the extra inductance in series with the capacitor.

From the circuit in Fig. 2.6, the differential mode voltage U_{DM} across the LISN resistor can be calculated. At twice the power line frequency (100 Hz or 120 Hz), capacitor C_F is selected to be a low impedance; therefore, it can be assumed that the capacitive reactance at the conducted transmit frequency (on the order of three or more) is close to zero. For example, when 50 kHz is an ideal 250 μF capacitor, the capacitive reactance is 0.01 Ω, and the parasitic L_F and R_F become the main impedance.

2.1.3 Equivalent Circuit of Conducted EMI Noise

1. Common Mode Conducted EMI Equivalent Circuit

Both the $U_L = U_N$ and $U_L \neq U_N$ results are obtained when testing the EUT's conducted EMI noise. Since the ratio of the noise voltage to the noise current is 50 Ω, the above two results can be expressed as $I_L = I_N$ and $I_L \neq I_N$. When $I_L = I_N$ or $I_L \neq I_N$, it is worth noting that there is always I_{DM} satisfied [4, 5]

$$I_L - I_{DM} = I_N + I_{DM} \tag{2.3}$$

where I_{DM} is the unbalanced noise current, which is the differential mode noise current. When $I_L = I_N$, I_{DM} is zero. When $I_L \neq I_N$, I_{DM} is not zero.

It can be known from Eq. (2.3) that regardless of $I_L = I_N$ or $I_L \neq I_N$, there is always I_{CM} satisfied

$$\begin{cases} I_{CM} = I_L - I_{DM} \\ I_{CM} = I_N + I_{DM} \end{cases} \tag{2.4}$$

In Eq. (2.4), I_{CM} is the balanced noise current; that is, after removing the unbalanced noise current in the live line and the neutral line noise, the remaining parts should be equal, so it is called the balanced noise current, that is, the common mode noise current.

It is worth noting that after the high-frequency noise current flows out from the live line, there are two kinds of loops, namely the live-ground line and the live-neutral line. Similarly, after the high-frequency noise current flows out from the neutral line, there are also two kinds of loops, namely the neutral-ground line and the neutral-live line.

Without loss of generality, assume that [5]

$$
\begin{aligned}
I_L &= I_{LG} + I_{LN} \\
I_N &= I_{NG} + I_{NL}
\end{aligned}
\tag{2.5}
$$

In Eq. (2.5), I_{LG}, I_{NG}, I_{LN}, and I_{NL} are high-frequency noise currents between the live-ground line, the neutral-ground line, the live-neutral line, the neutral-live line, respectively.

It is worth noting that the high-frequency noise currents between the live-neutral line and the neutral-live line are equal in magnitude and opposite in direction

$$
I_{NL} = -I_{LN}
\tag{2.6}
$$

Substituting Eq. (2.6) into Eq. (2.5)

$$
\begin{aligned}
I_L &= I_{LG} + I_{LN} \\
I_N &= I_{NG} - I_{LN}
\end{aligned}
\tag{2.7}
$$

Based on Eq. (2.7),

$$
\begin{aligned}
I_{LG} &= I_L - I_{LN} \\
I_{NG} &= I_N + I_{LN}
\end{aligned}
\tag{2.8}
$$

If $I_{LN} = I_{DM}$, then it can be obtained from Eqs. (2.4) and (2.8) that

$$
\begin{cases}
I_{CM} = I_L - I_{DM} = I_L - I_{LN} = I_{LG} \\
I_{CM} = I_N + I_{DM} = I_N + I_{LN} = I_{NG}
\end{cases}
\tag{2.9}
$$

Conclusion can be drawn from Eq. (2.9) that

$$
U_{CM} = U_{LG} = U_{NG}
\tag{2.10}
$$

where U_{CM} is the common mode conducted EMI noise. It can be seen from Eq. (2.10) that the conducted EMI noise between live-ground line and neutral-ground line is equal and the direction is the same, so it is common mode conducted EMI noise [5].

Fig. 2.7 Transmission path of common mode conducted EMI noise

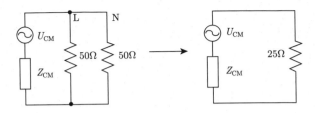

Fig. 2.8 Transmission path of differential mode conducted EMI noise

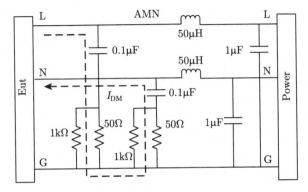

$$U_{\mathrm{CM}} = \frac{U_{\mathrm{L}} + U_{\mathrm{N}}}{2} \qquad (2.11)$$

From Eq. (2.11), the equivalent circuit of common mode conducted EMI noise is shown, as shown in Fig. 2.7, U_{CM} is the common mode noise source, and 25 Ω is the common mode LISN equivalent test impedance, which is formed with two parallel standard impedances of 50 Ω. Z_{CM} is the internal impedance of the common mode noise source, which needs to be further determined.

2. Differential Mode Conducted EMI Equivalent Circuit

From Eq. (2.10) and $I_{\mathrm{LN}} = I_{\mathrm{DM}}$, there is

$$I_{\mathrm{DM}} = \frac{I_{\mathrm{L}} - I_{\mathrm{N}}}{2} \qquad (2.12)$$

Since the ratio of noise voltage to noise current is 50 Ω,

$$U_{\mathrm{DM}} = \frac{U_{\mathrm{L}} - U_{\mathrm{N}}}{2} \qquad (2.13)$$

The transmission path and equivalent circuit of differential mode conducted EMI noise can be known from Eq. (2.13), as shown in Figs. 2.8 and 2.9. Z_{DM} is the common mode noise internal impedance and needs to be further determined.

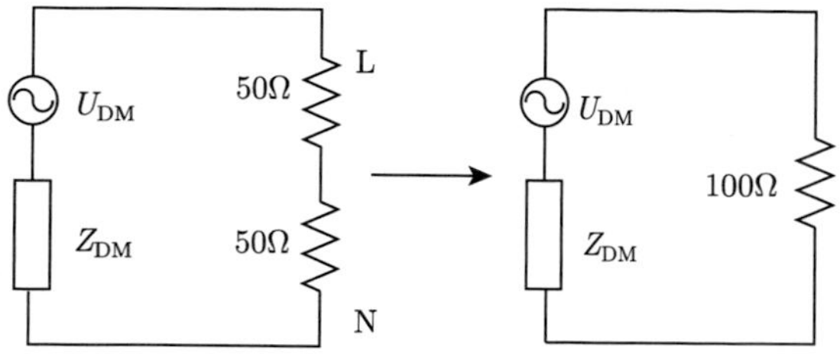

Fig. 2.9 Internal impedance of differential mode conducted EMI noise

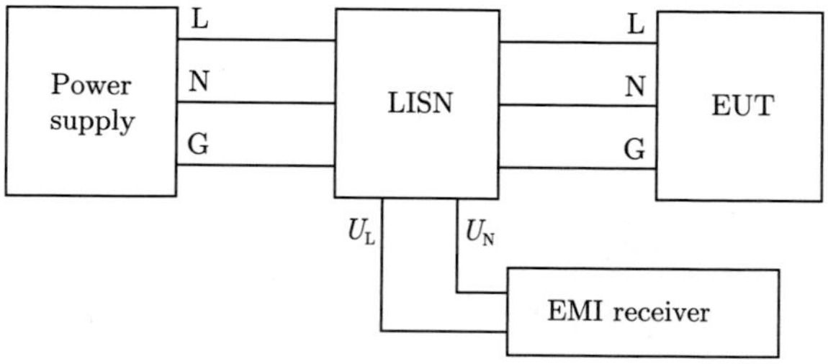

Fig. 2.10 Standard test method for conducted EMI noise

As shown in Fig. 2.9, U_{DM} is the differential mode noise source, $50\,\Omega$ is the differential mode LISN equivalent test impedance; that is, the standard impedance of two $50\,\Omega$ is formed in series, and Z_{DM} is the internal impedance of the differential mode noise source, which needs to be further determined.

2.2 Conducted EMI Noise Measurement Method

2.2.1 Line Impedance Stabilization Network Structure

Conducted EMI noise is measured by artificial mains network (artificial mains network, AMN), which is also called line impedance stabilization network (LISN). LISN is used to measure conducted EMI noise on electronic equipment power lines. The main function is to provide a stable impedance between the terminals of the device under test in the range 150 kHz–30 MHz of conducted EMI noise, the experimental circuit is isolated from the unwanted signal on the power supply to couple the interference voltage to the measurement receiving device. The test method is shown in Fig. 2.10. The equivalent impedance of the EMI receiver is $50\,\Omega$, and U_L and U_N are the conducted EMI noise of the live line and neutral line, respectively. The receiver processes the conducted EMI noise measured in the LISN and displays the noise spectrum on the computer.

The structure is shown in Fig. 2.11. For each power cord, the AMN is equipped with three ends: a power supplying terminal connected to the power supply, a testing terminal connected to the test device, and a noise outputting terminal connected to the noise measurement device.

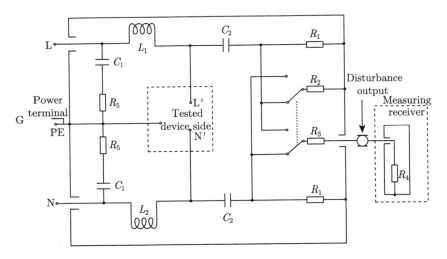

Fig. 2.11 LISN network topology

Table 2.1 Component values of the HV LISN

Element	Numerical value (Ω)	Element	Numerical value (μF)
R_1	5	C_1	8
R_2	10	C_2	4
R_3	1000	L_1	50
R_4	50	L_2	50
R_5	50 (receiver input impedance)		

   In this circuit, the inductors L_1 and L_2 are paths for the power signal and do not affect the power supply to the device under test. For high-frequency signals, it can be regarded as an open circuit, and the grid and the equipment under test are separated by high frequency. It can prevent high-frequency signals of grid from disturbing receiver. The capacitor C_1 disconnects the power signal from the receiver and is a path for high frequency interference from the device under test. L_1, C_1, R_1, R_3, and R_5 specify the impedance seen from the test end. The role of L_2, C_2, and R_2 is to prevent the influence of the unknown impedance of the grid on the impedance of the test end. The role of R_4 is to provide the correct end load for the other line while measuring one line. Component values of the HV LISN are shown in Table 2.1.

   For high-frequency signal flow, Fig. 2.11 can be abstracted into a three-port network as shown in Fig. 2.12.

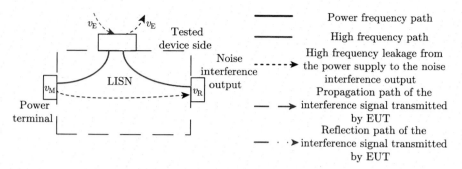

Fig. 2.12 Schematic diagram of the LISN three-port network

2.2.2 LISN Metrology Characteristic Parameters

For the current situation of absence of power supply impedance stabilization network, calibration specifications for the national standard GB/T 6113.102-2018 "Specification for radio disturbance and immunity measuring apparatus and methods—Part 1–2:Radio disturbance and immunity measuring apparatus—Coupling devices for conducted disturbance measurements" comprehensively and systematically describes the performance of the power supply impedance stabilization network. It summarizes the measurement characteristic parameters of the power supply impedance stabilization network, including two core parameters: the impedance of the test end, the partial pressure coefficient and an important parameter: isolation. From the theoretical point of view, the main performance indicators of the power supply impedance stabilization network are studied, which provides a theoretical basis for the subsequent calibration work.

1. Isolation Rejection Ratio (IRR)

According to GB/T 6113.102, in order to ensure that the unwanted signal on the power supply side and the unknown impedance of the power supply at all test frequencies do not affect the measurement, the EUT port-related terminal is terminated to a given terminal. Basic isolation (decoupling factor) requirements should be met between each power supply terminal and receiver port. This requirement applies only to the linear impedance stabilization network itself and does not include additional external cables and filters.

Definition Isolation rejection ratio refers to the logarithmic amplitude loss experienced by a high-frequency signal fed from the power supply and transmitted through the internal power supply network to the noise output (with the built-in attenuator not in the calculation range). If the voltage amplitude of the feed is U_p and the voltage amplitude measured at the noise output is U_m, the partial pressure coefficient is $20\lg(U_p/U_m)$. The transmission path contains a built-in attenuator.

When the high-frequency signal attempts to transmit from the power supply end of the LISN to the noise output end, the transmission path presents a large series high-

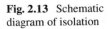

Fig. 2.13 Schematic
diagram of isolation

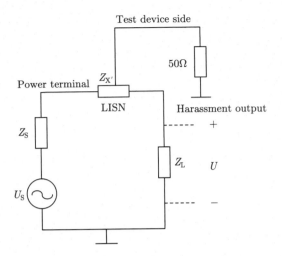

frequency impedance Z'_X due to the existence of the RLC isolation circuit (Fig. 2.13). In the figure, U_S is the electromotive force of the external excitation source, Z_S is the source internal impedance, and Z_L is the termination load impedance of the noise output. The Z'_X is much larger than the load Z_L (receiver inputting impedance), so the voltage value on the Z_L is small, which is equivalent to being isolated by the inductor. On the other hand, from the end of the device under test, the Z'_X and the device impedance Z_{EUT} of the device under test are connected in parallel, and Z'_X hardly affects the value of Z_{EUT} of the device end impedance of the device under test. When the power supply end of the LISN is connected to the grid, the unknown impedance of the grid connected to the Z'_X does not affect the value of the impedance of the device end Z_{EUT}, thus ensuring that the unknown impedance of the grid does not affect the impedance of the test end. Isolation is both voltage isolation and impedance isolation.

According to GB/T 6113.102, the $50\,\Omega/50\,\mu HV$ linear impedance stabilization network has a minimum isolation rejection ratio of 40 dB in the range of 0.15–30 MHz.

2. Voltage Division Factor

Definition The voltage division factor refers to the logarithmic amplitude loss experienced by the high-frequency signal fed from the device under test and transmitted through the AMN to the noise output (the built-in attenuator within the calculation range).

If the voltage amplitude fed is U_{out} and the voltage amplitude measured at the noise output is U_m, the voltage divider coefficient is $20\lg(U_{out}/U_m)$.

When the high-frequency noise of the device under test is transmitted from the test end along a line to the noise output, it can be represented by Fig. 2.14. The voltage division factor is U_{out}/U_m. In the figure, U_S is the electromotive force of the

Fig. 2.14 Schematic diagram of voltage division factor analysis

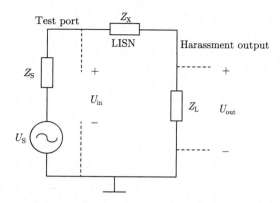

external excitation source, Z_S is the source internal impedance, Z_X is the equivalent impedance of the LISN transmission path, and Z_L is the termination load impedance of the noise output. Since the measurement result of conducted electromagnetic interference is generally expressed by dB μV, the voltage division factor can also be expressed as $20\lg(U_{out}/U_m)$.

The measured voltage division factor is used to compensate for the result of the modified conducted interference emission. Therefore, the voltage division factor has no allowable value range, and only the calibration laboratory is required to give the uncertainty.

3. EUT Port Impedance

Definition When the noise output end of a line is terminated with a precision load-50 Ω, the high-frequency impedance of the line terminal corresponding to the test end relative to the reference ground is called the impedance of the equipment under test (EUT Port Impedance). The impedance is in the plural form, including the mode and phase angle, or the real and imaginary parts.

The impedance values are specified by normalizing the values of the modes and phase angles. In the national standard, there is a clear tolerance requirement for the modulus and phase angle of the impedance, the allowable range of the modulus is ±20%), and the allowable range of the phase angle is ±11.5. It should be noted that the modulus of the impedance is required in the old standard GB/T 6113—1995, but the phase angle is proposed in the new standard.

4. Coupling Rejection Ratio

In addition to the above three indicators, when measuring a certain line, it is considered that not only the interference signal from the power grid but also the interference signal from other lines affect the interference output result. Therefore, large coupling should be avoided between the lines of the linear impedance stabilization network.

Definition The degree to which the receiver port signal is affected by the terminal signal of the uncorrelated device is the coupling rejection ratio.

With reference to isolation requirements, the coupling rejection ratio should be less than −40 dB.

2.2.3 Calibration of Measurement Characteristic Parameters

1. Calibration Method of the Impedance of the Test End

Disconnect the power supply from the linear impedance stabilization network calibrated and place it on the grounded conductive plate for good grounding. Set the sweep frequency range of the impedance measurement system to be the same as the frequency band of the calibrated, linear impedance stabilization network set the source power 0 dBm, and set the IF bandwidth 1 kHz. Connect the coaxial end of the dedicated calibration adapter to the impedance measurement system. On the non-coaxial end of the adapter, perform an "open-short-match" calibration for "single pin-ground" according to Fig. 2.15.

Perform the impedance calibration arrangement according to Fig. 2.16 to measure the power line impedance of the calibrated pins. The noise output of the network to be tested shall be terminated with a 50 Ω matching load. In addition, at the end of the device under test, the line other than the power LINE to be tested shall be terminated with a 50 Ω load. Also note that the distance impedance measurement system of manual power supply network is at least 1 m away and should not be located near and at the top to improve measurement repeatability.

The test frequency points are selected as required, and the impedance at each frequency, including the modulus and phase angle, are read by the impedance measurement system and recorded in the data sheet.

Change the power supply terminal that is to be tested (the device under test and the noise output should change at the same time), repeat the above steps until the ground impedance of all power terminals is measured.

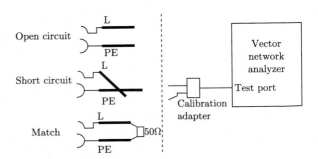

Fig. 2.15 "Open-short-match" calibration

Fig. 2.16 Arrangement of impedance calibration

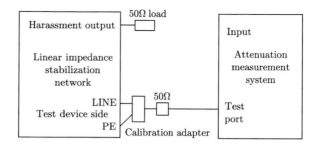

2. Calibration of the Voltage Division Factor

Disconnect the power supply from the linear impedance stabilization network and place it on the grounded conductive plate for good grounding.

Connect the test system according to Fig. 2.17, which is based on a vector network analyzer. The reference measurement is performed on the unknown characteristic of the impedance of the test end, and the purpose of the reference measurement is to obtain the actual input level of the test end. Note that the T-type route should be as close as possible to the end of the device under test [5, 6]. At the same time, each LINE of the linear impedance-stabilized network power supply terminal should be terminated with a 50 Ω load relative to the reference ground.

According to the operating frequency range of the network, the starting frequency and the ending frequency of the network analyzer are set. The measurement function selects the transmission measurement, the format is logarithmic amplitude, the source power level is set to -10 dB, and the intermediate frequency bandwidth is set 100 Hz or less. Then perform a pass-through calibration. Change the connection mode of the test system according to Fig. 2.17, and read the measurement results of the transmission coefficient at different frequency points, which is the measured value of the voltage division factor.

3. Calibration of IRR

Connect the test system according to Fig. 2.18. The north side of the power supply end of the linear impedance stabilization network is connected to the output of the

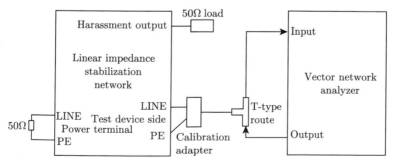

Fig. 2.17 Voltage division factor reference measurement

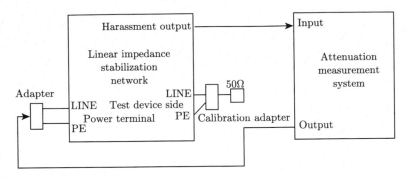

Fig. 2.18 IRR calibration measurement chart

attenuation measurement system through the adapter. The noise output is connected to the input of the attenuation measurement system. The corresponding LINE end of the device under test is terminated with a 50 Ω load on the reference ground. The measurement range of the attenuation measurement system is recommended to be smaller than 70 dB.

Select the test frequency point according to the requirements, and record the attenuation at different frequencies at this time, which is the IRR value of the linear impedance stable network to be tested at the frequency point.

Pay attention to the following aspects when calibrating IRR:

1. During calibration, the corresponding LINE of the device under test is terminated with a 50 Ω load on the reference ground, and the load termination has a greater impact on the calibration result of the isolation;
2. When a high-frequency signal is fed from the power supply terminal, there is a coaxial or non-coaxial RF adapter problem. It is recommended to use a normal adapter, but the shorter the length of the wire, the better.

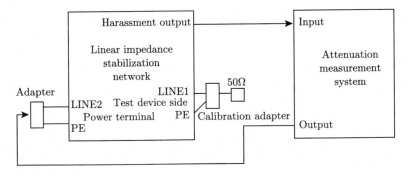

Fig. 2.19 Coupling suppression ratio calibration measurement chart

4. Coupling Suppression Ratio Calibration

When measuring a certain LINE, the noise signal of other lines can affect the measurement result of the LINE to be tested. The measurement method of the coupling suppression ratio is very similar to the IRR measurement. The high-frequency signal (level L_{ref}, logarithmic form) is fed from LINE 2 of the power supply terminal, and the receiver measures the output signal of LINE 1 at the noise output (level L, logarithmic form), then $(L - L_{ref})$ is the LINE 2 pairs of LINE 1 coupling suppression ratio. The specific measurement circuit is shown in Fig. 2.19. All LINE pairs on the device under test are terminated with a $50\,\Omega$ load.

2.3 Diagnosis of Conducted EMI Noise

2.3.1 Principles of Conducted EMI Noise Diagnosis

Since the measurement of conducted EMI noise is the total conducted EMI noise of the EUT, the conducted common mode noise and the conducted differential mode noise are not separated. However, the generation mechanism of conducted common mode noise and differential mode noise is also different, so there are also differences in the suppression methods. In order to effectively suppress conducted EMI noise, it is necessary to first diagnose it.

Figure 2.20 is a schematic diagram of a conducted EMI noise diagnosis system. After obtaining the EMI noise of the device under test (EUT) at the output of the main measuring device, it is input to the common mode CM/differential mode DM separation network for modal separation. The signals transmitted from the spectrum analyzer to the computer are then processed by the diagnosis software. The intelligent system not only provides independent common mode and differential mode

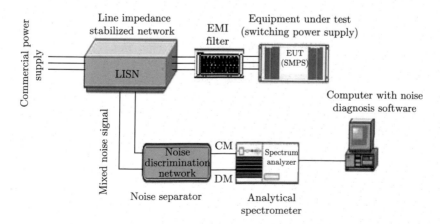

Fig. 2.20 Schematic diagram of a conducted EMI intelligent test system

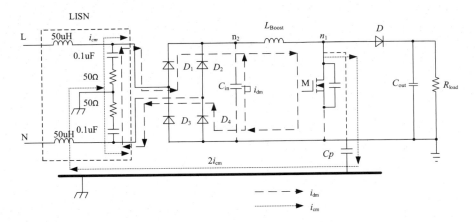

Fig. 2.21 Typical boost chopper circuit

components using hardware, but also provides useful information for filter design using software.

Conducted EMI can be classified into differential mode (DM) interference and common mode (CM) interference depending on the propagation path. Differential mode interference refers to the interference signal in the loop formed by the phase line and the neutral line of the power supply, and the common mode interference refers to the interference in the loop formed by the phase line or the neutral line of the power supply and the ground line.

Taking the boost chopper circuit (BOOST circuit) as an example, the differential mode (DM) interference and common mode (CM) interference propagation path and noise measurement method in power electronic circuits are analyzed. Figure 2.21 shows a typical boost chopper circuit.

1. Diagnosis of Differential Mode Conducted EMI

For a continuous conduction mode and a critical discontinuous mode PFC circuit, there is always a pair of diodes in the rectifier bridge at a certain time, such as D_1 and D_4, or D_2 and D_3 simultaneously turned on. The different propagation path of DM interference current is shown in Fig. 2.22.

The simplified differential mode interference equivalent circuit is shown in Fig. 2.23. This is a passive two-port network consisting of L_{BOOST} and C_m. The voltage gain $|U_0/U_n|$ can reflect the voltage transmission characteristics of the network. In the case of differential mode interference, the lower the voltage gain, the greater the suppression of the noise source by the network, and the smaller the interference signal received by the LISN. It can be seen that L_{BOOST} and C_{in} are the main factors affecting differential mode interference.

2. Diagnosis of Common Mode Conducted EMI

For systems where the heat sink is grounded, the parasitic capacitance C_m and LISN between the power tube and the heat sink form a common mode interference current

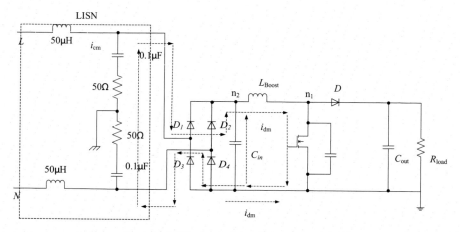

(a) Propagation path of difference mode interference current when D1 and D4 is conducted

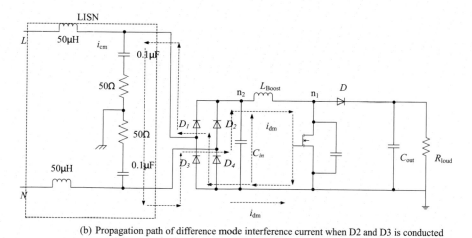

(b) Propagation path of difference mode interference current when D2 and D3 is conducted

Fig. 2.22 Differential mode interference current propagation path of BOOST circuit

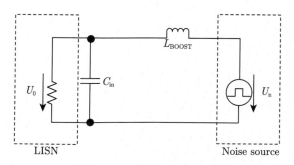

Fig. 2.23 Differential mode interference equivalent circuit of PFC circuit

path. Note that when the common mode current is balanced, the flow of the L and N lines is equal and the direction is the same; that is, the common mode current flowing through the ground is $2I_{CM}$.

According to Fig. 2.24, the common mode interference equivalent circuit can be obtained, as shown in Fig. 2.25. In the figure, I_{CM} is the common mode current of the ground, U_n is the noise source voltage, U_{CM} is the common mode voltage received by the LISN, and C_m is the parasitic capacitance between the drain of the switch and the heat sink.

It can be seen from the above analysis that when the switching frequency of the MOSFET changes, since the differential/common mode internal impedance does not substantially change, the noise does not change much.

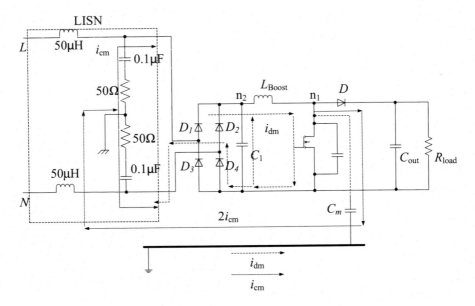

Fig. 2.24 Differential mode and common mode current in PFC circuit

Fig. 2.25 Common mode interference equivalent circuit

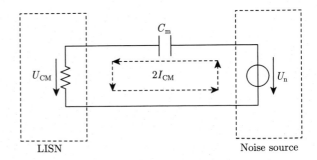

2.3.2   Noise Separation Network

Equations (2.11) and (2.13) show that the total noise of the live line and the neutral line contains common mode noise and differential mode noise, in which the common mode noise of the two lines is equal in size and phase. The difference mode noise in the two lines is equal in magnitude and opposite in phase [7].

$$U_L = U_{CM} + U_{DM}$$
$$U_N = U_{CM} - U_{DM} \tag{2.14}$$

1. Paul Network

The topology of the Paul network is shown in Fig. 2.26. It uses the RF transformer with a ratio of 1:1 as the core. As the phase switching of the neutral line noise, the switch can add or subtract the live line and the neutral line noise to realize the calculation process of Eqs. (2.11) and (2.13) by hardware.

2. See Network

The topology of the see network is shown in Fig. 2.27. The core transformer with a center tap ratio of 2:1 is used as the core. The center tap can be used to add or subtract the live line and neutral line to implement the calculations of Eqs. (2.9) and (2.11) in hardware.

Fig. 2.26 Paul network topology

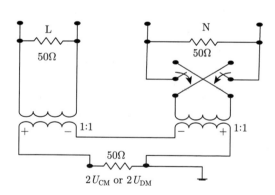

Fig. 2.27 See network topology

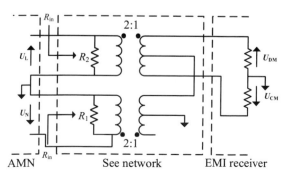

Although the see network eliminates the switches in the Paul network, like Paul, the mutual inductance generated between the two sets of RF transformers in the see network also affects the characteristics of the noise separation network.

3. Mardiguian Network

The topology of the Mardiguian network is shown in Fig. 2.28. Like the See network, it also uses a RF transformer with a center tap ratio of 2:1. But the center tap is in the primary coil, and the two sets of RF transformers are simplified into one group.

In engineering applications, both the live and neutral lines flow through the primary coil of the RF transformer. At the center of the primary coil, 1/2 live line noise and 1/2 neutral line noise can be added [7, 8]. At this time, the differential mode noise cancels each other out, and only common mode noise is shown in Eq. (2.15).

$$I_C = \frac{I_L + I_N}{2} = I_{CM} \tag{2.15}$$

On the other hand, the live line noise passes through the RF transformer, producing 1/2 of the neutral line noise on the secondary coil; the neutral line noise passes through the RF transformer, producing $-1/2$ neutral line noise on the secondary coil [8]. At this time, the common modes cancel each other out, only the differential mode, as shown in Eq. (2.16).

$$I_{\text{secondary coil}} = \frac{I_L - I_N}{2} = I_{DM} \tag{2.16}$$

As shown in Eqs. (2.15) and (2.16), the Mardiguian network can also implement the calculation process of Eqs. (2.11) and (2.13) by hardware.

The Mardiguian network abandoned the switch and only used a set of RF transformers to improve network performance. However, due to the distributed capacitance between the primary and secondary coils of the RF transformer, common mode noise flows into the secondary coil through the distributed capacitance, thereby degrading the characteristics of the Mardiguian network.

Fig. 2.28 Mardiguian network topology

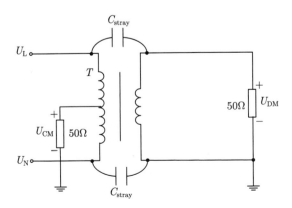

4. Guo Network

The topology of the Guo network is shown in Fig. 2.29. The structure is very simple. Only 0° and 180° power splitters are used. The 0° power splitter is also called a power combiner, which can directly add two sets of input signals and use the added half as the output, that is, common mode conducted EMI noise; the 180° splitter directly subtracts the two sets of input signals and use the subtracted half as the output, which is the differential mode conducted EMI noise.

However, since the power splitter has lower characteristics than the RF transformer in the frequency range of 9–30 kHz, and the hardware cost is high; therefore, the Mardiguian network is significantly better than the Guo network in terms of network characteristics and economics.

5. New Network Designed by the Author of This Book

Based on the Mardiguian network, considering the influence of distributed capacitance between RF transformers, the author of this book designed a new network, as shown in Fig. 2.30. It adds a common mode choke to the secondary coil of the

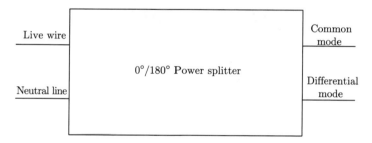

Fig. 2.29 Guo network topology

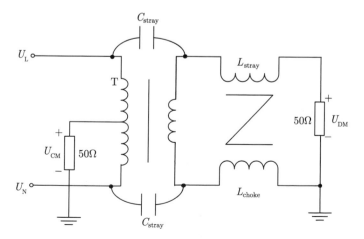

Fig. 2.30 New network topology

Mardiguian network to suppress common mode noise caused by the distributed capacitance of the RF transformer to improve network performance.

2.3.3 Noise Separation Network Characteristics Measurement Method

The noise separation network characteristics include the following four parameters.

(1) Common mode insertion loss CMIL: The common mode signal $U_{CM\text{-}in}$ is used as the input signal of the network to measure the common mode output signal $U_{CM\text{-}out}$ of the network. The CMIL is

$$CMIL = 20\lg \frac{U_{CM\text{-}out}}{U_{CM\text{-}in}} \tag{2.17}$$

In the formula, the CMIL ideal case is zero.

(2) Common mode rejection ratio CMRR: The common mode signal $U_{CM\text{-}in}$ is used as the input signal of the network and measuring the differential mode output signal $U_{DM\text{-}out}$ of the network, the CMRR is

$$CMRR = 20\lg \frac{U_{DM\text{-}out}}{U_{CM\text{-}in}} \tag{2.18}$$

In the formula, the CMRR ideal case is infinite.

(3) Differential mode insertion loss DMIL: The differential mode signal $U_{DM\text{-}in}$ is used as the input signal of the network, and the differential mode output signal $U_{DM\text{-}out}$ of the network is measured. The DMIL is

$$DMIL = 20\lg \frac{U_{DM\text{-}out}}{U_{DM\text{-}in}} \tag{2.19}$$

In the formula, the DMIL ideal case is zero.

(4) Differential mode rejection ratio DMRR: Using the differential mode signal $U_{DM\text{-}in}$ as the input signal of the network and measuring the common mode output signal $U_{CM\text{-}out}$ of the network, the DMRR is

$$DMRR = 20\lg \frac{U_{CM\text{-}out}}{U_{DM\text{-}in}} \tag{2.20}$$

In the formula, the DMRR ideal case is infinite.

In order to measure the characteristics of the noise separation network, including CMIL, CMRR, DMIL, and DMRR. The noise separation network characteristic test method is shown in Fig. 2.31.

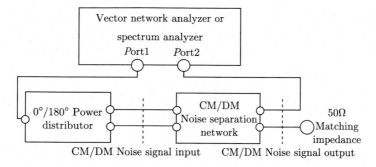

Fig. 2.31 Test method for noise separation network characteristics

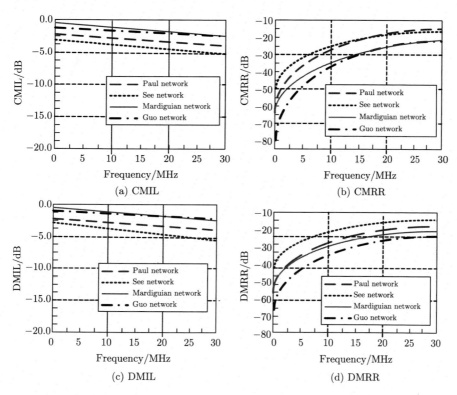

Fig. 2.32 Comparison of Paul, see, Mardiguian and Guo networks

As shown in Fig. 2.31, all ports in the test should be matched with 50 Ω impedance. The characteristics of the above four networks are tested by the Guwei GSP-827 spectrum analyzer, as shown in Fig. 2.32.

As shown in Fig. 2.32, the Mardiguian network has relatively good characteristics in the above four noise separation networks. However, in the Mardiguian network, the primary and secondary coils of the RF transformer have distributed capacitance

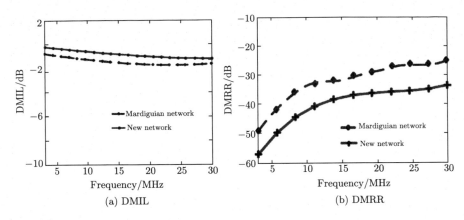

Fig. 2.33 Comparison of features between Mardiguian and the new network

through which common mode noise flows into the secondary coil, affecting the characteristics of the Mardiguian network. Since the main difference between the Mardiguian network and the new network is DMIL and DMRR, the above method was used to measure the Mardiguian network and the new network, as shown in Fig. 2.33.

As shown in Fig. 2.33, because the new network is based on the Mardiguian network, a set of common mode chokes is added to the secondary coil, which can effectively suppress high-frequency noise caused by parasitic capacitance coupling between the primary coils of the RF transformer. Therefore, the CMIL and CMRR characteristics of the new network are the same as those of the Mardiguian network, but the DMIL and DMRR characteristics of the new network are significantly better than the Mardiguian network. Therefore, the new network designed by the research group can improve the extraction accuracy of conducted EMI noise.

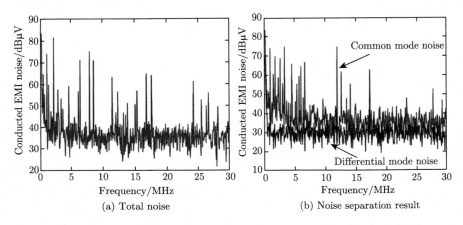

Fig. 2.34 Diagnostic test of EMI noise of switched reluctance motor (load power is 1 kW)

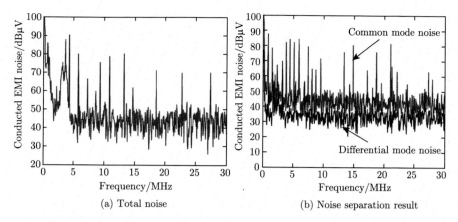

Fig. 2.35 Diagnostic test of EMI noise of switched reluctance motor (load power is 2.5 kW)

2.3.4 Experimental Verification

Taking a switch reluctance motor as an example, the new network is used to diagnose the conducted EMI noise. The switched reluctance motor has a rated voltage of 220 V and a rated power of 3 kW. The load during the test was 1 kW and 2.5 kW.

As shown in Figs. 2.34 and 2.35, the conducted EMI noise generated by the switched reluctance motor with load of 1 and 2.5 kW is dominated by common mode noise, which is consistent with practical engineering applications. Therefore, the noise separation network can effectively diagnose the mechanism of conducted EMI noise, extract common mode and differential mode conducted EMI noise, and provide theoretical basis for subsequent conducted EMI noise suppression.

References

1. Fluke JC. Controlling conducted emissions by design. IEEE Electromagn Compat Mag. 1991;4(4):48–9.
2. Tihanyi L. Electromagnetic compatibility in power electronics. Piscataway, NJ: IEEE Press; 1995.
3. Morrison R. Grounding and shielding techniques in instrumentation. 3rd ed. Hoboken: Wiley; 1986.
4. Henry WO. Noise reduction techniques in electronic systems. 2nd ed. Hoboken: Wiley; 1988.
5. Nave MJ. Power line filter design for switched-mode power supplies. 2nd ed. Hoboken: Wiley; 1991.
6. Ozenbaugh RL, Pullen TM. EMI filter design. New York: Marcel Dekker Inc.; 2000.
7. Zhao Y. Noise diagnosis techniques in conducted electromagnetic interference (EMI) measurement: methods analysis and design. In: Antennas and propagation society international symposium; 2004.
8. Zhao Y, See KY. Performance study of CM/DM discrimination network for conducted EMI diagnosis. Chin J Electron. 2003;12(4):536–8.

Chapter 3
Conducted EMI Noise Suppression Methods and Case Study

3.1 Suppression Principle of Conducted EMI Noise

3.1.1 Internal Noise Suppression

Taking the switching power supply as an example, according to the analysis model of differential mode and common mode, the generation and propagation of the conductive electromagnetic interference of the switching power supply are related to the driving circuit noise and the circuit parameters. The following is an analysis of the switching power supply interference mechanism from the spectrum of the driving noise and the circuit parameters.

Switching devices for switching power supplies are now using more MOSFETs and IGBTs. They can generate relatively large voltage and current rate of change when turned off. When the switching device in the switching power supply is turned off, the rate of change of voltage/current is large, causing great interference. In order to suppress the interference of the switching power supply, it is necessary to understand the spectral characteristics of the noise signal generated by the interference source. Also, it is necessary to study the influence of the rise/fall time of the signal on spectral characteristics.

The driving signals of some switching devices are trapezoidal signals, but such signals do not exist in practice. When designing a circuit, in order to improve the response speed, a drive signal with a short rise and fall time is generally selected. These signals can cause more interference. These noises can affect the circuit through parasitic parameters in the circuit. In order not to lose generality, the asymmetric ladder signal is first studied [1–3].

The asymmetric ladder signal is shown in Fig. 3.1, and the amplitude spectrum formula of the signal is [2–4]

$$S(k) = \frac{A}{\pi k} \left| \sin(\pi k R) e^{j\pi k d} - \sin(\pi k F) e^{-j\pi k d} \right| \tag{3.1}$$

© Science Press 2021
Y. Zhao et al., *Electromagnetic Compatibility*,
https://doi.org/10.1007/978-981-16-6452-6_3

Fig. 3.1 Asymmetric ladder signal

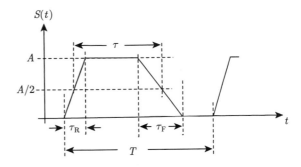

$si(x) = \sin(x)x$, $R = \tau_R/T$, relative rise time $F = \tau_F/T$, relative fall time $d = \tau/T$, and relative average pulse width is duty cycle.

In order to avoid overlapping of two pulses, d, R, and F must be satisfied, $d + R + F \leqslant 1$ and $R + F \leqslant d/2$.

Analyze Eq. (3.1) as follows:

(1) Low-frequency band

For the fundamental signal, it can be approximated as [4, 5]

$$S(1) = \frac{A}{\pi}\sqrt{[\sin(\pi R) + \sin(\pi F)]^2 \sin^2(\pi d) + [\sin(\pi R) - \sin(\pi F)]^2 \cos^2(\pi d)}$$

$$(3.2)$$

For the low-frequency band, the rise time and fall time can be neglected, that is, $R, F \to 0$, then $\sin(\pi kR) \approx \sin(\pi kF)$, so the low-frequency range amplitude spectrum can be approximated as [5]

$$S(k)_{LF} = \frac{2A}{\pi k} \tag{3.3}$$

The amplitude of the spectrum in the low-frequency band is proportional to the amplitude of the trapezoidal wave.

(2) Middle-frequency band

The mid-band amplitude spectrum can be approximated as

$$S(k) = \frac{A}{\pi k}\left(1 + \frac{1}{\pi k \alpha}\right), \quad \alpha = \begin{cases} R & \text{when } R < F \\ F & \text{when } F < R \end{cases} \tag{3.4}$$

If $R > F$, the amplitude of the spectrum is proportional to the amplitude of the trapezoidal wave and inversely proportional to R.

(3) High-frequency band

At high frequencies, $|\sin(\pi k R)| \leqslant |1/\pi k R|$, $|\sin(\pi k F)| \leqslant |1/\pi k F|$, Eq. (3.4) can be approximated as [5]

$$S(k)_{HF} = \frac{A}{\pi^2 k^2} \left(\frac{1}{R} + \frac{1}{F} \right) \tag{3.5}$$

The amplitude of the spectrum at this time is proportional to the amplitude of the trapezoidal wave and inversely proportional to R.

In summary, when the frequency is constant, as the rise time increases, R increases, and the amplitude of the spectrum in the middle- and high-frequency bands decreases, resulting in a reduction in EMI noise. Therefore, paralleling the proper capacitance between the base and ground can slow the rising edge and increase the rise time, thus achieving the effect of reducing the radiated noise.

3.1.2 External Noise Suppression

1. Principle of Common Mode Noise Suppression

Analysis was performed with conducted EMI of the PFC converter. Figure 3.2 shows a schematic diagram of the CRM boost PFC circuit with the LISN circuit added.

Consider the switch drain–source voltage U_{DS} as a common mode conduction noise source, the common mode noise mechanism of the boost PFC circuit is analyzed. The LISN structure is simplified and shown as two $50\,\Omega$ test impedances. In the figure, C_p represents the parasitic capacitance of the drain of the MOS transistor, and i_p is the current flowing through the parasitic capacitance. The analysis process is divided into positive and negative half cycles of the input voltage. As shown in Fig. 3.3a, when the bridge diodes D_1 and D_4 are turned on, the input voltage is in the positive half cycle, and the noise source U_{DS} is applied to the parasitic capacitor C_p to continuously charge and discharge to form a current i_b, which flows through the test resistor to the ground and forms a complete loop through diode D_4.

Similarly, when the rectifier diodes D_2 and D_3 are turned on, the input voltage is in the negative half cycle, and the current i_p generated on the capacitor flows through the LISN test resistor to the ground and forms a complete loop through the diode D_3.

It can be seen from Fig. 3.3 whether it is the positive half cycle or the negative half cycle, that the noise source U_{DS} generates a common mode current through the parasitic capacitance C_p to the ground and flows through the test impedance of the L line and the N line through the ground, and the LISN impedance can be seen as $25\,\Omega$ at this time, which is a parallel connection of two $50\,\Omega$ resistors [6]. To reduce the common mode conducted emission, the controllable parameters are the parasitic capacitance C_p and the mixed power line voltage U_{in}.

In practical applications, common mode rejection is typically achieved by common mode capacitors between the ground and ground, and the line-to-ground capacitance is typically half the maximum allowed for leakage requirements. From the

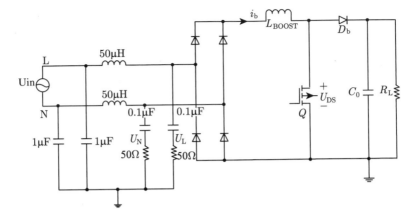

Fig. 3.2 Critical boost PFC converter circuit

perspective of the common mode conduction voltage, two line-to-ground capacitances are seen in parallel; therefore, the equivalent common mode capacitance is equal to the sum of the two capacitance values.

2. Principle of Differential Mode Noise Suppression

Differential mode interference is caused by high-frequency interference currents in the loop, while differential mode conducted interference of the PFC converter is generally considered to be caused by the ripple of the inductor current i_b. Figure 3.4 shows the differential mode EMI path analysis [7]. Like the common mode noise analysis, the differential mode conducted EMI analysis is also divided into positive and negative half cycles. As shown in Fig. 3.4a, when the diodes D_1 and D_4 in the rectifier bridge are turned on, the differential mode interference current generates equal valued voltages in opposite directions on the two-line test resistor R_{LN}.

As shown in the figure, whether it is a positive half cycle or a negative half cycle, a current is driven by a switching transistor along a loop at a switching frequency. This loop is mainly composed of a switching transistor and a rectifier bridge LISN test resistor. For differential mode currents, the LISN can be seen as $100\,\Omega$, which is the series connection of the L and N two-wire test impedances.

For differential mode noise, the two Y capacitors need to be connected in series. Therefore, the equivalent differential mode capacitance is only equal to half the value of one capacitor. This only provides very little differential mode filtering, especially at the low frequency that is most needed, and the capacitance value cannot be increased because of leakage requirements. These capacitors contribute to differential mode attenuation above about 10 MHz, which is usually not required [8, 9]. Therefore, they are usually ignored for differential mode filtering.

In order to provide a large-capacity differential mode capacitor, a line-to-line capacitance X capacitor is necessary for the power line filter. Because this capacitor is not connected to ground, the value is not limited by the leakage requirements. Typical values for this capacitor range from 0.1 to 2 μF. For safety reasons, a resistor

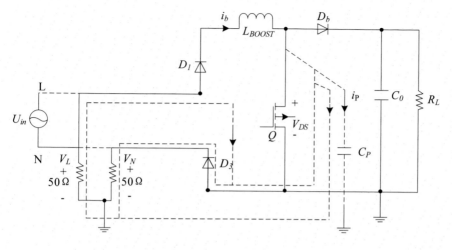

(a) Positive half-cycle common mode EMI path analysis

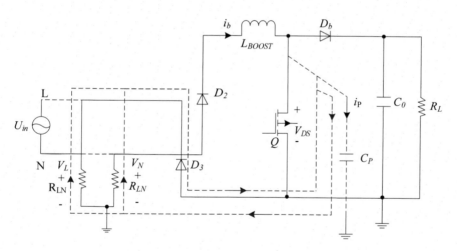

(b) Negative half-cycle common mode EMI path analysis

Fig. 3.3   Common mode EMI path analysis

(usually 1 MΩ) is sometimes added in parallel with this capacitor. This resistor is used to discharge the capacitor when the power is turned off.

When the power supply has a poor-quality ripple filter capacitor, or two capacitors are connected in series, it is helpful to set a second X capacitor on both sides of the power line and on the power supply side of the common mode choke.

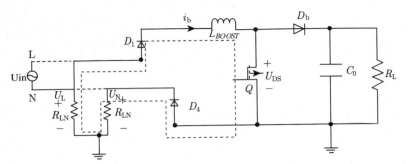

(a) Positive half-cycle differential mode EMI path analysis

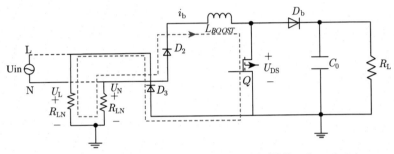

(b) Negative half-cycle differemtial mode EMI path analysis

Fig. 3.4 Differential mode EMI path analysis

3.2   Suppression Methods of Conducted EMI Noise

3.2.1   Ground

1. Power Ground

Safety grounding is different from the purpose of signal grounding. It is the outer case of the electrical equipment connected to the ground with a low-impedance conductor, which reduces the chance of electric shock hazard if it is accidentally touched. Z_1 in Fig. 3.5a is the leakage resistance between the case and the point where the potential U_1 is located, and Z_2 is the leakage resistance between the case and the ground. The potential of the case is determined by the resistance of Z_1 and Z_2, so the case potential is [9, 10]

$$U_K = \left(\frac{Z_2}{Z_1 + Z_2} \right) U_1 \qquad (3.6)$$

At this time, the case potential may be quite high, and the value depends on the size of Z_1 and Z_2. If Z_1 is much larger than Z_2, the potential of the case is close to the value of U_1, and there is a danger of electric shock.

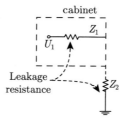

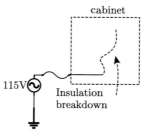

(a) The chassis potential depends on the size of Z_1 and Z_2

cabinet

115V

Insulation
breakdown

(b) Fused AC power line is introduced into the enclosed enclosure

Fig. 3.5　Case should be grounded for safety to avoid insulation breakdown

If the case is grounded, that is, $Z_2 = 0$, it can be known from Eq. (3.7) that the case potential U_K should be 0 V. Currently, if a person touches a grounded case, since the impedance of the human body is much larger than 0, most of the current flows into the ground through the grounding wire, so there is no problem with electric shock.

Figure 3.5b shows a more dangerous situation showing the introduction of a fused AC power line into a closed enclosure.

If the power line touches the case, the case can provide the current that the fuse can withstand outside the case. If a person touches the case, the current of the power line enters the ground directly through the human body. If the grounding measures are implemented, when an insulation breakdown or power line touches the case, a large amount of current flows on the power line due to the grounding to burn the fuse, so that the case is no longer charged, and there is no danger of electric shock.

2. Signal Ground: Single-Point and Multi-point Grounding

Another typical location is the signal ground, which allows the signal current to return to the source. Although designers want signals to return to the source through the path they are designed for, there is no guarantee that this can happen. In fact, some spectral components of one signal are returned to the source through one path, while other spectral components of the same signal may be returned to the source through another path. A shielded cable on the ground plane is a good example. Spectral components below the cutoff frequency of the shielded ground circuit return along the ground plane, while those spectral components above the cutoff frequency return along the shield rather than return along the ground plane. In the case of signal ground, it is important to consider the path through which the current flows. Again, the high-frequency impedance of these paths is primarily inductive, and if they "ground" with a small voltage drop across them, then this inductive impedance must be minimized.

There are basically two options for signal grounding: single-point grounding systems and multi-point grounding systems. A single-point grounding system means that the ground loop of the subsystem is only connected to a single point within the subsystem. The purpose of using a single-point grounding system is to prevent currents in two different subsystems from sharing the same loop return, resulting in common-impedance coupling [10]. Figure 3.6 shows the implementation principle of a typical single-point grounding. The three subsystems have the same signal source. The method shown in Fig. 3.6a is called "cascade chain" or tandem method. This approach clearly produces a common-impedance coupling between the ground points of the two subsystems.

The connection method in Fig. 3.6a adds the signals in SS#2 and SS#3 to SS#1, as discussed earlier. The portion of the underlying fold line is the loop of the current that must be known. When they are determinable, the parallel connection shown in Fig. 3.6b is the ideal single-point grounding method. However, it has a big disadvantage, that is, the impedance of a single ground wire depends on the length of these wires. In a distributed system, if the principle of a single-point grounding system is strictly obeyed, the connecting line may take a long time. In this way, the ground lines can have a large impedance and eliminate their positive effects. Moreover, the loop currents on these wires are likely to radiate effectively to other ground conductors and create coupling between subsystems, like crosstalk, thus creating radiated emission compliance issues. The extent to which this occurs depends on the spectral components of the loop signal: The high-frequency components can produce more efficient radiation and coupling than the low-frequency components. Therefore, the single-point grounding principle is not a universally applicable ideal grounding principle because it is best suited for low-frequency subsystems.

Another type of grounding system is the multi-point grounding system shown in Fig. 3.7a. Typically, a large conductor (usually a ground plane) acts as a signal loop in a multi-point grounded system. In a multi-point grounding system, and each ground of the subsystem is connected to a ground conductor at a different point. The reason why a multi-point grounding system is superior to a single-point grounding system is that the length of the connecting wire may be shorter because there is a closer grounding point. But this again assumes that the impedance between the ground points at the frequency is zero or at least very low, which is not correct. If the ground plane shown in Fig. 3.7a is replaced by a long and narrow strip line on the PCB, then if the ground of the subsystem is connected along each point on the strip line, then it can be considered as multi-point grounding. In fact, this is more like the series connected single-point grounding system shown in Fig. 3.6a.

Another problem with multi-point grounding systems may be that other currents through the ground conductor are not noticed. For example, assume that a "ground plane" (subsystem is multi-point grounded) intentionally has other currents or ambient currents passing through it. As illustrated in Fig. 3.7b, the same as the PCB of other digital circuits includes a DC motor drive circuit. The +38 V DC power required to drive the DC motor and the +5 V DC power required to energize the digital cir-

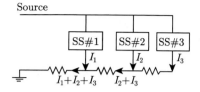

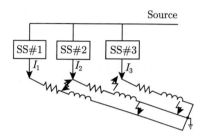

(a) Common impedance coupling in a series connection

(b) Unintentional coupling between ground and single-point grounding systems

Fig. 3.6   Single-point grounding problem description

Fig. 3.7   Example of multi-point grounding

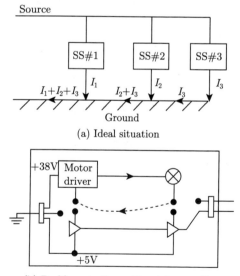

(a) Ideal situation

(b) Problems in the multi-point grounding

cuit are supplied to the PCB through the connector. Assume that these circuits are grounded on a common ground on the PCB. As the motor drives switch, the high-frequency current of the motor circuit passes through the ground plane, and a large high-frequency voltage is generated between the two points of the ground grid. If the digital logic circuit is also connected to the grounding grid in multiple points, the voltage generated by the motor loop current on the grounding grid may couple into the digital logic circuit causing problems in performance. In addition, if a signal on the PCB passes through a power connector on the opposite side of the PCB, the ground conductor in the signal cable is excited by the varying potential of the noisy grounding system, possibly generating radiation and causing radiation or conduction emission problems.

Typically, a single-point grounding system is used in the analog subsystem, including low-level signals. In these cases, a ground voltage drop of millivolts or even microvolts can cause significant common-impedance coupling interference problems in these circuits. Single-point grounding systems are also commonly used in high-level subsystems, such as motor drives, to prevent these high-level loop currents from creating large voltage drops across the common ground plane. On the other hand, digital subsystems are inherently resistant to external noise; however, they are quite sensitive to internal noise, and interference caused by internal noise through common-impedance coupling can be considered as "shooting from the heart to itself" in order to this kind of common-impedance coupling is minimal [10, 11]. The grounding system in the digital system tends to be grounded at multiple points. The large ground plane is used, such as the ground in the inner layer of the PCB or a grounding grid, to place many alternate ground paths in parallel, thus reducing impedance of the loop. It is important to place signal conductors near the ground loop as this also reduces the impedance of the loop.

Other types of grounding systems are called hybrid grounding systems, where the two systems are combined in different frequency bands. As an example, Fig. 3.8 has alternative ways of achieving shield grounding and avoiding low-frequency coupling. If the cable contains two shield layers, the inner shield is connected to one end of the reference conductor, and the outer shield is connected to the other end of the reference conductor, then there is no low-frequency connection between the two shields, which avoids the reference conductor. The problem of common-impedance coupling is caused by the current flowing upward. However, the parasitic capacitance between the two shield layers (the parasitic capacitance is concentric due to the two shield layers) is quite large, providing a high-frequency connection between the two shield layers so that the shield layer can effectively interact with the reference conductor connected to the end. This is the frequency selective grounding mechanism of the hybrid grounding system. The capacitor shown in Fig. 3.9a provides a single-point grounding system at low frequencies and a multi-point grounding system at high frequencies. The inductance shown in Fig. 3.9b is just the opposite. The grounding scheme shown in Fig. 3.9b is useful when it is necessary to connect the subsystem to the ground for safety and to ground at a higher frequency.

A typical system requires two or three separate grounding systems. As shown in Fig. 3.10a, the low-level signal (voltage, current, power) subsystem should be con-

Fig. 3.8 Mixed grounding
scheme in actual device

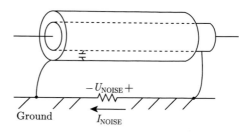

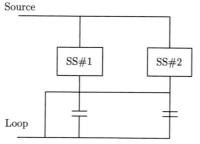

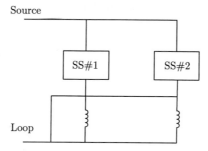

(a) Single-point grounding at low frequencies and multi-point grounding at high frequencies

(b) Single-point grounding at high frequencies and multi-point grounding at low frequencies

Fig. 3.9 Mixed grounding scheme

nected to a dedicated separate ground point, which is the signal ground. In this signal ground subsystem, the circuit uses single-point grounding, multi-point grounding or hybrid grounding. The second type of grounding system refers to a noise grounding system. The noise grounding system represents a circuit that operates at a high level and/or produces a noise-like signal. In one case, the signal is noise, and in another case, it is not. For example, the high spectral components of a digital clock signal are considered to be noise when they meet specified limits or interfere with other subsystems, although they are necessary spectral components of the useful signal. On the other hand, the brush arc of a DC motor is noisy, and it is not necessary for the function of the motor. A PCB containing digital circuits, analog circuits and noise, motor drive circuits is shown in Fig. 3.10b. The ground of the noise circuit

Fig. 3.10 Independent
grounding system

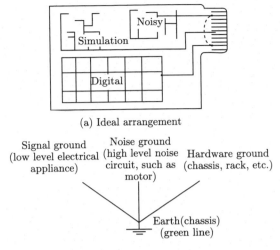

(a) Ideal arrangement

(b) Layout of the grounding system on the PCB

is specifically connected to the PCB connector to prevent high-level return current from passing through the analog or digital grounding system. Similarly, digital and analog circuits also have a dedicated ground loop connected to the connector. Note that the ground in the analog grounding system (one signal ground) is essentially a single-point grounding system, while the grounding in the digital grounding system (the other signal ground) is essentially a multi-point grounding system.

The key to understanding why these different and independent grounding systems are needed is that their purpose is to prevent common-impedance coupling. If high-level noise is allowed to be transmitted from the motor drive circuit to the conductor that is the digital circuit loop, these high-level currents create a voltage drop across the common circuit provided to the digital circuit, which may cause problems with the function of the digital circuit. It is important to distinguish between low and high loops because the greater the magnitude of the loop current, the greater the voltage drop across the common impedance. Several different low-level circuits may share the same loop without interfering with each other because the common-impedance coupling voltage drop generated on the common ground is not large enough to cause interference. Not only are signal levels important in separate grounding systems, but the spectrum is also important. Some branches contain intrinsic filtering at their inputs, so if the noise spectrum falls outside the passband of the circuit's input filtering, the high-level noise signal applied to the input cannot cause interference problems. Digital circuits tend to have very wide wideband inputs so there is no frequency selective protection. On the other hand, an analog circuit such as a comparator has a certain degree of high-frequency filtering due to the response time of the operational amplifier. However, parasitic phenomena can weaken this effect. Hardware grounding is usually separated from other groundings to avoid common-impedance coupling problems. The 60 Hz high-level signal, like the ESD signal, can

Fig. 3.11 Grounding of a system with cabinet

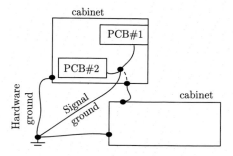

pass through this ground plane. It is important not to provide a connection between the hardware ground and other ground, especially the signal ground, such that the voltage drop produced by the ESD signal transition does not produce a point of noise variation within the signal ground system.

3. Hardware Ground

It refers to connecting between the bottom plate, the base, the case, the equipment rack, etc. The purpose of hardware grounding is not to carry current for use in the event of a fault or the transfer of ESD signals.

Consider the situation shown in Fig. 3.11, which is an example of a problem that may arise after a separate grounding system is connected. Two cabinet or brackets form the system, and their hardware grounding is always connected to a common grounding point. These enclosures are also connected to prevent potential generation between them due to, for example, ESD (electrostatic discharge) current flow. Two PCBs in one enclosure connect their signal grounds together, but it is generally incorrect to connect the signal ground to the chassis because the ESD discharge described above may cause the signal ground to change with ESD discharge. However, in some cases it may be necessary to connect these grounds together to prevent ESD problems. The parasitic capacitance and parasitic inductance between the chassis and the internal circuitry can cause the high-frequency performance of the grounded system to deviate significantly from the ideal state. In this way, the circuits should be physically isolated from each other and isolated from the chassis enclosure to avoid high-frequency coupling.

4. Ground Loop Current Voltage and Ground Noise Suppression

There is a certain potential difference between two different grounding points called the ground voltage. This is because there is always a certain impedance between the two grounding points, the ground current flows through the ground common impedance, a ground voltage is generated thereon, and the ground voltage is directly applied to the circuit to form a common mode interference voltage. For example, in the ground loop shown in Fig. 3.12, current from a DC power source or a high-frequency signal source is returned via the ground plane. Since the common impedance of the ground plane is very small, it is often not considered in the perfor-

Fig. 3.12 EMI generated by common ground impedance

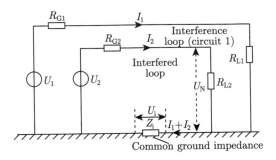

mance design of the circuit. However, for electromagnetic interference, the presence of ground plane impedance must be considered in the loop.

Therefore, there is a common impedance Z_i between the interference loop and the disturbed loop as shown in Fig. 3.12, and the voltage present on the common impedance is $U_i = U_1 + U_2 = Z_i I_1 + Z_i I_2$. For the disturbed loop, $Z_i I_1 = U_1$ is the electromagnetic interference voltage, and $Z_i I_2 = U_2$ is the partial pressure of the load voltage drop. Because of $R_{L2} \gg |Z_i|$, the effect of $Z_i I_2$ on the load voltage drop is generally ignored and only considers the effect of electromagnetic interference voltage caused by I on the load. If the effect of the current I_2 of the disturbed loop on the ground common impedance Z_i is not considered, that is, $U_2 = 0$, the current I_1 in the interference loop (circuit 1) generates an interference voltage $U_i = U_1$ on the ground common impedance Z_i, and this voltage drops. The load R_{L2} of the interfered loop is disturbed, and the interference voltage is [11]:

$$U_N = \frac{Z_i R_{L2}}{(R_{G1} + R_{L1}) \times (R_{G2} + R_{L2})} U_1 \tag{3.7}$$

The interference received by the load R_{L2} of the interference loop is a function of the interference source U_1 of the circuit 1.

(1) Formation of Ground Current and Ground Voltage

Electronic equipment generally uses a metal plate with a certain area as a grounding surface. For various reasons, there is always a grounding current passing through the grounding surface, and there is always a certain impedance between the two points of the metal grounding plate, thereby generating a grounding interference voltage. The existence of grounding current is the root cause of grounding interference, and the grounding current is mainly caused by the following four types.

The grounding current caused by the conductive coupling is in many cases using two-point grounding or multi-point grounding, that is, connecting to the ground plane through two or more points, thus forming a ground loop through which the grounding current flows, as shown in Fig. 3.13.

The grounding current formed by capacitive coupling, due to the distributed capacitance between the loop component and the ground plane, can form a ground loop through the distributed capacitor, and some current in the circuit always leaks into the

Fig. 3.13 Conductively coupled ground current loop diagram

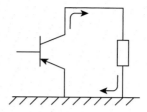

ground loop, as shown in Fig. 3.14. Figure 3.14a indicates a ground loop formed by a combination of conductive coupling and capacitance, and a ground current flows through the ground loop. Figure 3.14b indicates a ground loop formed by distributed capacitance at two points of high potential and low potential of the impedance element. When the loop is in a resonant state, the ground current becomes large.

Inductive ground current is formed by electromagnetic coupling, and when the coil in the circuit is close to the device housing, the housing is equivalent to a secondary coil having only one turn. It forms a transformer coupling with the primary coil, the ground current is generated by the electromagnetic induction in the case, and regardless of the position of the coil, an induced current is generated if the magnetic flux passes through the case.

The ground current is formed by the antenna effect of the metal conductor. When a radiated electromagnetic field is irradiated to the metal conductor, the conductor generates an induced electromotive force due to the receiving antenna effect. If the metal body is a box structure, the electric field acts on two parallel planes. A potential difference is generated to cause a ground current to flow through the case, and when the metal case is connected to the circuit, a current loop through which a ground current flows is formed.

When two devices connected by a transmission line are placed near the ground, as shown in Fig. 3.15, the common mode interference of the transmission line converts the external electromagnetic field into the common mode interference voltage in the loop. Although the loop impedance formed is high or low, the interference voltage induced in the loop is independent of the impedance, and the conductive coupling and the capacitive coupling are formed. The common ground impedance voltage is directly related to the loop impedance and current.

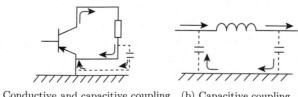

(a) Conductive and capacitive coupling (b) Capacitive coupling

Fig. 3.14 Capacitive coupled ground loop

Fig. 3.15 Common mode interference formed by external electromagnetic waves on the transmission line

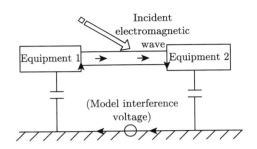

(2) Ground Loop Interference

It can be seen from the above analysis that the interference current and voltage are formed in the ground loop by the grounding common impedance and the antenna effect of the transmission wire or the metal case, and the interference voltage is induced to the input end of the victim circuit through various ground loops and the formation of ground loop interference.

Figure 3.16a shows the actual circuit of the ground loop interference, the common mode interference voltage U_i, and the flow interference current on the loops ABCDE-FGHA and ABCIJFGHA. Since the impedances of the two loops are different, the interference current is at the amplifier input. The terminal can generate an interference voltage. The transmission line and load shown in the figure can be balanced or unbalanced. The ground loop interference has different properties depending on the circuit structure and the type of transmission line. Figure 3.16b shows the equivalent circuit of ground loop interference.

Figure 3.17 shows the four isolation methods for eliminating ground loop measures.

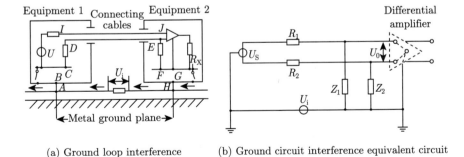

(a) Ground loop interference   (b) Ground circuit interference equivalent circuit

Fig. 3.16 Ground loop interference and equivalent circuit

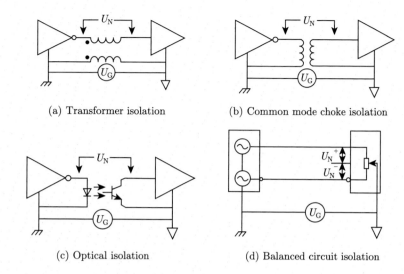

(a) Transformer isolation   (b) Common mode choke isolation

(c) Optical isolation   (d) Balanced circuit isolation

Fig. 3.17   Measures to eliminate the ground loop

3.2.2   Conducted EMI Noise Suppression Devices

1. Common Mode Choke

The common mode choke can disconnect the ground loop when DC or low frequency is continuous to suppress conducted current and voltage on the ground loop. As shown in Fig. 3.18, the performance of this circuit can be analyzed with an equivalent circuit. The signal source voltage U_S is connected to the load R_L via the connection line resistors R_{C1}, R_{C2}, and the longitudinal choke is represented by the inductances L_1, L_2 and the mutual inductance M. If the two coils are identical and form a tight coupling on the same core, then $L_1 = L_2 = M$, U_G is the longitudinal voltage of the ground loop that is magnetically coupled or formed due to the ground potential difference. Since R_{C1} and R_L are connected in series and $R_{C1} \ll R_L$, R_{C1} is omitted.

First analyze the response of the circuit to U_S. The equivalent circuit of U_S is shown in Fig. 3.19. It is known that when the frequency is higher than $\omega = 5R_{C2}/L_2$, all of the current I_S returns to the signal source through one of the following wires without

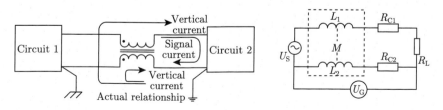

Fig. 3.18   Equivalent model of choke at DC or low frequency

Fig. 3.19 Equivalent circuit of U_S

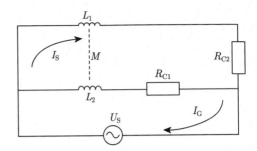

Fig. 3.20 Equivalent circuit of U_G

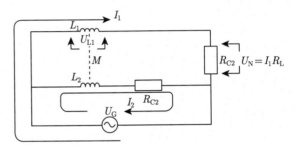

flowing through the ground line. If L_2 is selected, at the lowest signal frequency, $\omega > 5R_{C2}/L_2$, $I_G = 0$, then, in the loop below, there is

$$U_S = j\omega(L_1 + L_2)I_S - 2j\omega M I_S + (R_L + R_{C2})I_S \qquad (3.8)$$

Because $L_1 = L_2 = M$, there is $I_S = U_S/(R_L + R_{C2}) \approx U_S/R_L$.

Since $R_L \geqslant R_{C2}$, and the transformer is equal to no access, that is, when the transformer inductance is large enough and the signal frequency is higher than $5R_{C2}/L_2$, the addition of the transformer has no effect on the transmission of the signal.

See Fig. 3.18 for the response to the vertical voltage U_G. The equivalent circuit at this time B is shown in Fig. 3.20. The noise voltage is applied to both ends of the R_L when no transformer is added. If a transformer is installed, the noise voltage applied to the R_L can be calculated by the following two loops:

I_1 Loop:

$$U_G = j\omega L_1 I_1 + j\omega M I_2 + I_1 R_L \qquad (3.9)$$

I_2 Loop:

$$U_G = j\omega L_2 I_2 + j\omega M I_1 + I_2 R_{C2} \qquad (3.10)$$

$$I_2 = \frac{U_G - j\omega M I_1}{j\omega L_2 + R_{C2}} \qquad (3.11)$$

Because $L_1 = L_2 = M = L$, and based on Eq. (3.11), there is

Fig. 3.21 Relationship between U_N/U_G and frequency

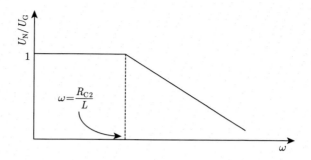

Fig. 3.22 Winding coils on the same magnetic ring

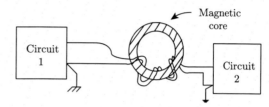

$$I_1 = \frac{U_G R_{C2}}{j\omega L(R_L + R_{C2}) + R_{C2} R_L} \tag{3.12}$$

Because $U_N = I_1 R_L$, $R_{C2} = R_L$, by Eq. (3.13), there is:

$$U_N = \frac{U_G R_{C2}/L}{j\omega + R_{C2}/L} \tag{3.13}$$

Figure 3.21 shows the relationship between U_N/U_G and frequency. If you want to reduce the noise voltage, you should reduce R_{C2} as much as possible, and the inductance of the transformer should be

$$L = \frac{R_{C2}}{\omega} \tag{3.14}$$

where ω is the angular frequency of the noise. It should be noted that the core section of the transformer should be large enough to not saturate when there is a certain amount of unbalanced DC current.

The longitudinal choke coil or neutralization transformer of Fig. 3.22 is relatively simple to manufacture. It can be wound with two wires on one magnetic ring. The longitudinal choke coil is arranged in a simple method of winding with two wires (axle cables can also be used to replace the wires), which does not form crosstalk between the circuits. In a telephone device, typically 25–50 circuits can be connected to a typical neutralization transformer.

2. Capacitor

In power equipment, the role of capacitors is mainly energy storage, smoothing, blocking and decoupling, and different types of capacitors, loss characteristics and frequency characteristics are relatively large, suitable for different occasions. At present, common types of capacitors generally include paper capacitors, ceramic capacitors, safety capacitors, and thin film dielectric capacitors. Among them, the ceramic capacitor is small, the equivalent series resistance at high frequency is small, and the performance of the safety capacitor does not cause electric shock and does not endanger personal safety. These two capacitors are used more in the anti-jamming circuit, often used in combination with the inductor, or in parallel with the circuit to bypass the high-frequency noise signal. As shown in Fig. 3.23, the conducted EMI suppression principle of the capacitor under ideal conditions can be regarded as the effective power supply of the circuit and the series connection of the noise power supply U_{CM}/U_{DM}, and the capacitance C can be the high-frequency noise current I_{CM}/I_{DM} in the circuit. Meanwhile, it does not affect the normal power supply to achieve conducted EMI suppression.

In capacitance suppression principle in the low-frequency state, the capacitor is generally considered to have ideal capacitance. As shown in Fig. 3.23, the amplitude of the ideal impedance of the capacitor decreases linearly with the frequency. However, the characteristics of the capacitor at high frequencies deviate far from the ideal state.

The actual capacitor is composed of an insulation leakage resistance R_p, an equivalent series resistance R_S, an equivalent series inductance L_p, and a capacitor. The capacitor leakage resistance R_p is caused by the capacitor material itself, the capacitor lead leads to the equivalent series resistance R_S, the capacitor lead and structure lead to the equivalent series inductance L_p, and it is the inductance component that influences the capacitance frequency characteristic in high frequency. The equivalent circuit of the actual capacitor is shown in Fig. 3.24.

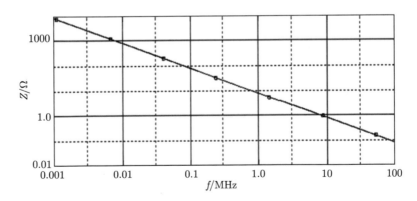

Fig. 3.23 Capacitor ideal impedance

Fig. 3.24 High-frequency equivalent circuit of actual capacitor

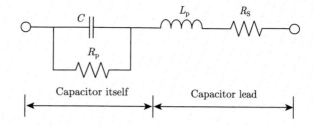

High-frequency capacitor impedance can be expressed as

$$Z_C = R_S + j\omega L_p + \frac{R_p}{1 + j\omega L_p C} \qquad (3.15)$$

According to Eq. (3.16), the measured impedance of the capacitor is shown in Fig. 3.25.

The capacitor exhibits a resistance characteristic at a very low frequency, and the value mainly depends on the insulation leakage resistance R_p caused by the material itself. When the frequency is gradually increased, the influence of the insulation leakage resistance R_p is reduced, which is approximately equivalent to

$$Z_C = R_S + j\omega L_p + \frac{1}{j\omega C} \qquad (3.16)$$

In the EMI filter design, the main consideration is the actual characteristics of the capacitor at higher frequencies. The insulation leakage resistance R_p can be ignored to form the high-frequency equivalent circuit model of the capacitor in Fig. 3.26. The simplified high-frequency model consists of a capacitor C, an equivalent series resistance R_S, and an equivalent series inductance L_p connected in series.

Fig. 3.25 Impedance plot of the capacitor

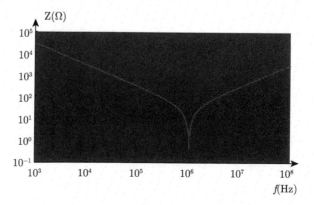

Fig. 3.26 High-frequency simplified equivalent circuit of actual capacitor

The frequency at which the impedance is minimized is called the resonance frequency f_r. When the low-frequency band $f < f_r$, the impedance of the capacitor is capacitive and inversely proportional to the frequency; when $f > f_r$, the impedance of the capacitor is inductive and proportional to the frequency; when $f = f_r$, the equivalent inductance L_p and the capacitor C series resonance occur, and the capacitor impedance Z_C is the smallest. At this time, the decoupling ability of the capacitor is the strongest. In actual engineering, it is necessary to find a capacitor that suppresses the noise closest to the resonant frequency and the noise frequency. The series resonant frequency can be expressed as

$$f_r = \frac{1}{2\pi\sqrt{L_p C}} \tag{3.17}$$

The high-frequency characteristics of the actual capacitor are primarily dependent on the equivalent series inductance L_p caused by the capacitor structure and leads. The inductance of the capacitor is typically 5–50 nH. The length of the lead and the resonant frequency is the main factor affecting the high-frequency characteristics of the capacitor. Therefore, in order to maximize the noise suppression capability, the external lead should be shortened during installation.

As shown in Fig. 3.27, L_p takes 5 nH, 10 nH, and 20 nH respectively. At 30~100MHz, the insertion loss of the capacitor is inversely proportional to L_p. The larger the insertion loss of the filter, the better the filtering effect is. Therefore, the extraction and suppression of the parasitic inductance are important to improve the filtering performance of the filter.

The parasitic inductance of the capacitor can be obtained by impedance measurement. It can be seen from the circuit that when series resonance occurs, the circuit becomes a pure resistor. Measurement of the resonant frequency, the parasitic inductance L_p, and the parasitic resistance R_S are as follows:

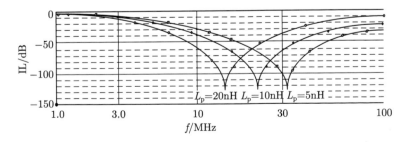

Fig. 3.27 Effect of parasitic inductance on capacitor insertion loss

Table 3.1 Resonant frequency and L_p of capacitor

C	f (MHz)	L_p (nH)	C (pF)	f (MHz)	L_p (nH)
1 μF	1.7	9	820	1.7	21
0.1 μF	4	16	680	4	21
0.01 μF	12.6	16	560	12.6	22
3300 pF	19.3	16	470	19.3	22
1800 pF	25.5	16	390	25.5	22
1100 pF	33	21	330	33	21

$$L_p = \frac{1}{\omega^2 C} \tag{3.18}$$

$$R_S = |Z_C| = \left| R_S + j\omega L_S + \frac{1}{j\omega C} \right|_{\omega = 2\pi f_r} \tag{3.19}$$

The parasitic inductance L_p of the capacitor is derived from the lead inductance and is estimated by Eq. (3.20).

$$L_p = 2l \left(\ln \frac{4l}{d} - 0.75 \right) \times 10^{-7} \tag{3.20}$$

where l is the lead length (cm) and is the d diameter (cm).

If the wire length is short ($l < 100d$), the perception of parasitic inductance tends to be extreme in the relatively high-frequency range.

$$L_{p\infty} = 2l \left(\ln \frac{4l}{d} - 1 \right) \times 10^{-7} \tag{3.21}$$

If Eq. (3.22) is used, there is a maximum error of 6% at around several hundred kHz frequencies and only 2% error at higher frequencies, which is completely allowed in engineering applications. Table 3.1 shows the resonant frequency of the ceramic capacitor with a lead length of 1.6 mm.

When the length of the lead is constant, the capacitor series resonant frequency is smaller as the capacitance value is smaller. Therefore, in order to suppress high-frequency noise, the series resonance of the capacitor itself should be suppressed as much as possible.

3.2.3 EMI Filter

1. Characteristics of EMI Filter

Filters are used in all areas of electronic engineering, such as communication, signal processing, and automation. There is a lot of design information about this type of filter. It is important to note that power supply filters used to reduce conducted emis-

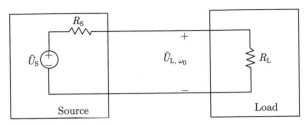

(a) Loaded voltage without filter

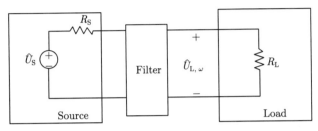

(b) Loaded voltage with filter

Fig. 3.28 Definition of filter insertion loss

sions are rarely designed using these traditional filter design methods. However, a discussion of the basic principles of traditional filters helps to discover the principles of all filters.

A typical characteristic of a filter is the insertion loss (IL), which is generally expressed in dB. Now consider the problem of providing a signal to the load as shown in Fig. 3.28a. To prevent certain frequency components of the source from reaching the load, insert a filter between the source and the load, as shown in Fig. 3.28b.

The loaded voltage without filter is $\hat{U}_{L,\omega_0}$, and the loaded voltage with filter is $\hat{U}_{L,\omega}$. The insertion loss of the filter is defined as follows:

$$\mathrm{IL}_{\mathrm{dB}} = 10\lg\left(\frac{P_{L,\omega_0}}{P_{L,\omega}}\right) = 10\lg\left(\frac{U_{L,\omega_0}^2/R_L}{U_{L,\omega}^2/R_L}\right) = 20\lg\left(\frac{U_{L,\omega_0}}{U_{L,\omega}}\right) \qquad (3.22)$$

Note that the voltage in this formula is not represented by the symbol (^), so it only represents the voltage amplitude. Due to the insertion of the filter, the input loss reduces the loaded voltage at a certain frequency. Usually, the insertion loss appears as a function of frequency.

The insertion loss of the low-pass filter is shown in Fig. 3.29a. The loaded voltage without a filter can be easily obtained from Fig. 3.28a.

$$\hat{U}_{L,\omega_0} = \frac{R_L}{R_S + R_L}\hat{U}_S \qquad (3.23)$$

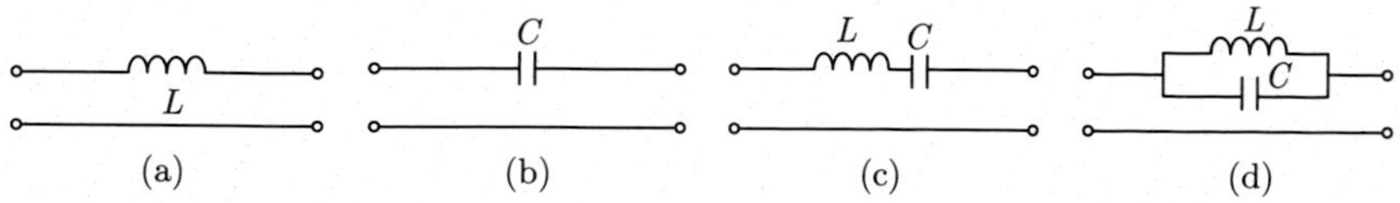

(a)   (b)   (c)   (d)

(a) Low pass   (b) High pass   (c) Hand pass   (d) Hand stop

Fig. 3.29 Four simple filters

Loaded voltage with filter

$$\hat{U}_{L,w} = \frac{R_L}{R_S + j\omega L + R_L}\hat{U}_S = \frac{R_L}{R_S + R_L}\frac{1}{1 + j\omega L/(R_S + R_L)}\hat{U}_S \qquad (3.24)$$

IL is

$$IL = 20\log_{10}\left|1 + \frac{j\omega L}{R_S + R_L}\right| = 20\log_{10}\left[\sqrt{1 + (\omega\tau)^2}\right] = 10\log_{10}[1 + (\omega\tau)^2] \tag{3.25}$$

Which τ is,

$$\tau = \frac{L}{R_S + R_L} \tag{3.26}$$

It is the time constant of the circuit. The insertion loss curve is from 0 dB of DC to the frequency component $\omega_{3\,dB}$, and other higher-frequency components decay very quickly. The expression of the insertion loss at frequencies above 3 dB is simplified to

$$IL \approx 10\log_{10}[(\omega\tau)^2] = 20\log_{10}(\omega\tau) = 20\log_{10}\left(\frac{\omega L}{R_S + R_L}\right) \quad \omega \gg \frac{1}{\tau} \tag{3.27}$$

Analyses of other filters are similar.

The above example illustrates an important point: The insertion loss of a filter depends on the impedance of the source and the load and therefore cannot be given independently of the termination impedance. Most filter manufacturers provide a frequency response curve for filter insertion loss. Since, the insertion loss depends on the impedance of the source and load. Assume $R_S = R_L = 50\,\Omega$ to obtain the impedance values of the source and load. This brings up a problem that how the insertion loss index affects filter performance based on $50\,\Omega$ source and load impedance work in conducted emission testing. The load impedance corresponds to the $50\,\Omega$ impedance of the LISN between the phase line and the ground line, between the neutral line and the ground line. However, in a typical configuration, $50\,\Omega$ is doubtful when the impedance seen from the grid is R_L. So, what is the source impedance R_S? The answer is unclear because the source impedance needs to be seen from the product's power input, which is $50\,\Omega$ and is constant over the frequency range of the conducted emission test. Therefore, using the insertion loss data provided by the manufacturer to evaluate the performance of the filter in the product may not achieve the desired result.

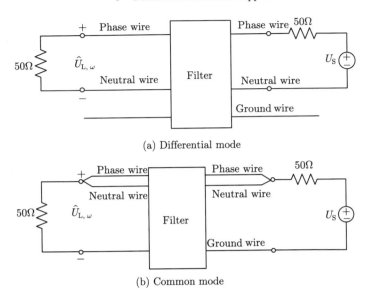

Fig. 3.30 Insertion loss measurement

There are two types of current that must be reduced: common mode current and differential mode current. Filter manufacturers typically give insertion loss for different currents. The data are obtained as shown in Fig. 3.30. As shown in Fig. 3.30a, the ground wire end differential mode insertion loss is measured. The ground line is not connected. The phase wire and the neutral wire form the circuit under test. Since, the differential mode current is defined as flowing out of the phase line and returning through the neutral line. So, there is no differential mode current on the ground line. For the common mode current test, the phase and neutral lines are connected to form a test circuit with a ground line as shown in Fig. 3.30b. Again, assume that the source and load impedance are both 50 Ω.

2. LC Filter Design

LC filter is a filter circuit composed of a combination of inductor, capacitor, and resistor. It can filter out one or more harmonics. The most common LC filter structure is the inductor. In series with the capacitor, it can form a low-impedance bypass for the main subharmonics (3, 5, 7); single-tuned filters, double-tuned filters, and high-pass filters are all LC filters. The LC filter is designed to be very low impedance at a certain frequency, and the corresponding frequency harmonic current is shunted. The behavior mode is to provide a passive harmonic current bypass channel. The basic form of LC filter circuit is shown in Fig. 3.31.

The design of LC filter parameters should meet the following principles: ① The relationship between capacitor C, inductor L, and resistor R satisfies the filtering principle of the LC filter; ② the fundamental equivalent impedance of the LC filter meets the requirements of system reactive power compensation; ③ The overall

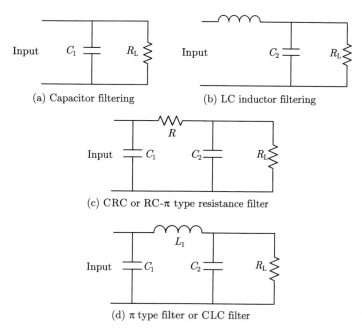

(a) Capacitor filtering   (b) LC inductor filtering

(c) CRC or RC-π type resistance filter

(d) π type filter or CLC filter

Fig. 3.31 Basic form of LC filter circuit

impedance of the LC filter should not form a series or parallel resonance with the grid impedance; ④ The harmonic content of the power grid after installing the passive filter is lower than the national standard; ⑤ The calculation of the harmonic capacity of each group of passive filters must include not only the harmonic capacity filtered by each, but also the background harmonic capacity of 10%.

Filters are used to change the characteristics of a signal or, in some cases, to cancel a signal. The filter can be differential or common mode. The differential mode filter is well understood, and there are many books and papers on the design. However, common mode filters are often considered mysterious and poorly understood.

Common mode filters are typically used to suppress noise in the cable when the expected differential mode signal is allowed to pass without interference. Why is a common mode filter more difficult to design than a differential mode filter? There are basically three reasons:

(1) The source impedance is unknown;
(2) The load impedance is unknown;
(3) The filter does not distort the useful signal (differential mode signal) on the cable.

The effectiveness of a filter depends on the source and load impedance with the filter in between. For a differential mode filter, it is often easy to find information about the driver's output impedance and load impedance. However, for common mode signals, the source is the noise generated by the circuit (not specified where), the load is usually some cable as an antenna, and the impedance frequency is usually unknown, and changes with frequency, cable length, conductor diameter, and layout of cable.

For common mode filters, the source impedance is usually the ground impedance of the printed circuit board (PCB) (because it is inductive, so small and increases with frequency), and the load is the cable impedance of the antenna (except that the cable accessory resonance is large). Although the exact impedance of the source and load is unknown, there are methods to deal with the amplitude and frequency characteristics.

For the common mode filters, in order to not distort the useful differential mode signal, the differential mode of the filter must generally satisfies the following conditions:

(1) For the narrowband signal, the highest frequency occurs;
(2) For wideband digital signals, the frequency of the signal $1/\pi t_r$ (t_r is the rise time).

Differential mode or signal line filters (e.g., clock line filters, etc.) should be placed as close as possible to the source or driver. However, the common mode filter should be placed as close as possible to the location of the cable in and out of the enclosure.

Figure 3.32 shows a simple two-element, low-pass, common mode filter consisting of a series element and a bypass element. The filter is embedded in the signal conductor and the return conductor. The figure also shows the common mode (noise) and differential mode (signal) voltage sources connected to the filter.

For a common mode voltage source, the two bypass capacitors are connected in parallel to the total capacitance of $2C_{shunt}$. For differential mode voltage sources, the two capacitors are connected in series to the total capacitance $C_{shunt}/2$. Therefore, the common mode source faces 4 times the capacitance of the differential mode source. This result is good because the effect on the common mode signal of capacitor filter is expected to be more significant than the effect on the differential mode signal.

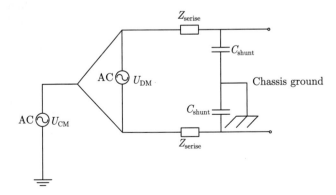

Fig. 3.32 Two-cell common mode filter with signal and return conductors

However, for a common mode source, the two series impedances are connected in parallel to the total impedance. For differential mode sources, the two series impedances are connected in series to 2 of the total impedance. Therefore, the differential mode source faces four times the impedance of the common mode source. This result is not good because the series impedance of the filter should have more effect on the common mode voltage than on the differential mode voltage. As a result, the series elements in the common mode filter are often constructed as common mode chokes. In this case, the differential mode impedance is zero, and the series impedance only affects the common mode signal without affecting the differential mode signal.

A common mode filter is usually a low-pass filter consisting of one–three components in one of the following topologies:

(1) Single component filter;
(2) A single series element;
(3) A single bypass element.

Multi-element filter:

(1) L-type filter (one series element and one bypass element);
(2) T-type filter (two series elements and one bypass element;
(3) π-type filter (two bypass elements and one series element).

The advantage of a single component filter is that they require only one component. The advantage of a multi-element filter is that they are effective when a single component filter is ineffective and provides more attenuation than a single component filter.

The bypass component in the filter is almost always a capacitor, the value of which is determined by the effective frequency range of the filter. The series component can be a resistor, an inductor, or a ferrite. If the DC voltage drop is acceptable, a resistor can be used; if the DC voltage drops unacceptable, apply inductors or ferrites, both of which have zero or very small DC voltage drops, apply inductors at low frequencies ($<10<30$ MHz), and ferrites at high frequencies. Sometimes a small resistor can be added in series with the inductor to reduce Q value, which is also an additional advantage that the ferrite structure of a common mode choke does not affect the differential mode.

Filter attenuation occurs because of impedance mismatch. From the above discussion, it is known that the source impedance is usually very low, and the load impedance is usually very high except for resonance. Therefore, L-type filters use low-impedance components (series components) to face low source impedance, while low-impedance components (bypass capacitors) should be most effective in facing high load impedance.

For the series impedance filter component to be effective, there must be an impedance greater than the sum of the source and load impedance. For the bypass filter component to be effective, there must be a parallel combination of impedance less than the source and load impedance.

Therefore, the three possible scenarios are as follows: (1) Source and load impedance are low. In this case, the series component is effective; (2) Source and load impedance are high. In this case, the bypass capacitor is effective; (3) source or one of the load impedances is low, and the other is high. In this case, no single component filtering is effective, and a multi-element filter must be used.

Near the cable resonance (both source and load impedance), the series components are most effective, and when the resonant frequency is higher than the cable, the bypass capacitor is most effective (medium to high source impedance, high load impedance). The cable impedance drops to a low point with multiple resonance points, at which point the series components are active. In the case of a cable as a dipole antenna, there can be an impedance of approximately 70 Ω for the first resonance point and approximately 35 Ω for a monopole antenna of the cable.

About multistage filters, the more stages of the filter, the less the attenuation depends on the termination impedance, and most of the mismatch occurs between the components of the filter itself, independent of the actual source and load impedance.

The bypass capacitor needs to have a low-impedance connection to ground. When controlling the common mode noise generated by the PCB ground, the filter capacitor needs to be connected to the chassis or chassis ground. If, as recommended, the circuit ground is connected to the chassis ground in the input/output (I/O) area of the PCB, the chassis ground and circuit ground are the same at that point.

Note that the common mode filter needs to be applied to all conductors that leave or enter the equipment enclosure, including the circuit ground conductor. When using a bypass capacitor, each capacitor, including the ground conductor, must be connected by a capacitor. In the case of series resistance or inductance, one component must be placed in series with each conductor including the ground conductor. However, for a ferrite core, all the wires are passed through a ferrite core, and all the wires in the cable can be handled with one component. This is one of the main advantages of ferrite: One component can handle many wires.

When a series element in the L-type filter is used with a bypass capacitor, the series element should be placed on the side of the filter circuit (because this is the low-impedance side) and the capacitor is placed on the cable side (because this is a high-impedance side). This structure is required because the circuit of the low source impedance is grounded and the capacitor does not work. If a high-impedance series component is placed on the circuit side of the bypass capacitor, the series component effectively increases the source impedance to a value that makes the capacitor effective. If ferrite is placed on the cable side of the capacitor, the original high cable impedance increases with little effect on the filter.

3. EMI Filter Design

Filters made in the field of electromagnetic compatibility are different from general-purpose high-pass, low-pass, band-pass, and band-stop filters. The common filter and EMI filter use environment, and filter indicators are completely different. The rated working current of the EMI filter is large, the working voltage is high, and it is required to withstand the impact of instantaneous large current. The characteristics of

Fig. 3.33 Insertion loss of
the filter

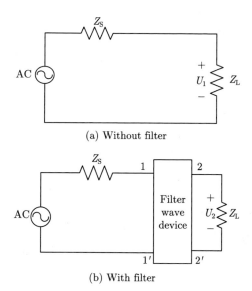

(a) Without filter

(b) With filter

the load end and the characteristics of the source end are greatly different depending on the working environment.

The filter has the following important technical parameters: rated current, rated voltage, insertion loss, leakage current, operating temperature rise, etc. The most important indicator for evaluating filter performance is insertion loss.

The insertion loss (IL) of a filter is a function of frequency. The unit is decibel (dB), which is the ratio of the power transmitted to the load before and after the filter is connected to the circuit. Let P_1, P_2 and U_1, U_2 be the values of the load power and voltage when it is with and without filter. Let P_1, P_2 and u_1, u_2 be the values of the load power and voltage when it is with and without filter as shown in Fig. 3.33. Z_L is the load impedance, Z_S is the source impedance of the input, and the insertion loss formula is

$$\mathrm{IL} = 10 \lg \left(\frac{P_1}{P_2} \right) = 10 \lg \left(\frac{U_1^2/Z_L}{U_2^2/Z_L} \right) = 20 \lg \left(\frac{U_1}{U_2} \right) \tag{3.28}$$

Generally, the EMI filter proposed in the study only suppresses conducted interference, that is, reduces the noise in the frequency band of 150 kHz to 30 MHz. The commonly used EMI conduction filter is wound by a choke coil and an enameled wire and is equipped with an X capacitor and a Y capacitor at the input end and the output end. Conducted EMI noise includes common mode conducted noise and differential conducted mode noise. Correspondingly, the conducted EMI filter also includes a common mode conducted EMI filter, a differential mode conducted EMI filter, and a common/difference mode conducted EMI filter. The typical topology is shown in Fig. 3.34.

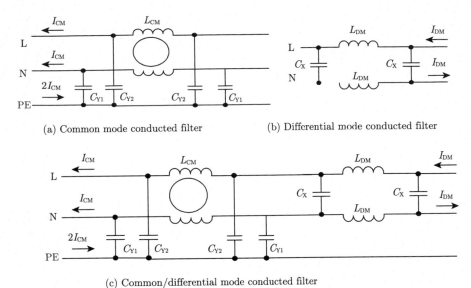

(a) Common mode conducted filter   (b) Differential mode conducted filter

(c) Common/differential mode conducted filter

Fig. 3.34 Conducted filter typical topology

The common mode conducted EMI filter consists of a common mode choke and a capacitor. The typical topology is shown in Fig. 3.34a. It is mainly used for system cables such as DC power lines and ACs that cause common mode EMI noise due to ground reflection.

The differential mode EMI filter consists of differential mode inductors and capacitors. The typical topology is shown in Fig. 3.34b. It is mainly for system cables with large loop areas, such as sensor cables in medical electronics.

Common/differential mode mixed EMI filters are often used on power lines. The typical topology is shown in Fig. 3.34c. This filter combines the advantages of common mode and differential mode filters and is mainly installed on AC power cable, so the conduction part of the cable integrated EMI filter uses a common/differential mode hybrid topology model.

Conventional conducted filter only suppresses conducted noise, and there is almost no suppression of radiated noise. Reducing the noise current in the electronic system plays a key role in EMI noise suppression, and suppressing the noise current in the AC power cable is a more effective and feasible method. Based on the conducted EMI filter installed on the AC cable, the cable radiated EMI filter module is added to form a power cable integrated conduction/radiation EMI filter, as shown in Fig. 3.35.

The noise current I_{AC} in the power cable can be expressed as

$$I_{AC} = I_{AC-CE} + I_{AC-RE} \tag{3.29}$$

I_{AC-CE} indicates the conducted interference current on the power cable, and I_{AC-RE} indicates the radiated interference current on the power cable.

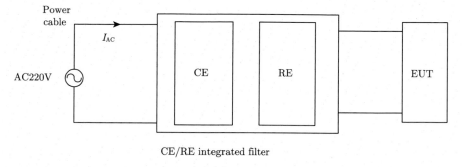

Fig. 3.35 Conducted/radiated integrated filter

The total frequency band of EMI test is 150 kHz–1 GHz, in which the conducted interference frequency test section is 150 kHz–30 MHz, and the radiated interference test frequency band is 30 MHz–1 GHz. Detailed structural optimization of the conducted/radiated integrated filter. The conducted EMI filter module follows a common/differential mode conducted filter. The radiated noise frequency band is divided into three sections: The low-frequency radiating section is 30–100 MHz, and the capacitor is formed by a capacitor with similar conduction; the intermediate frequency radiating section is 100–400 MHz, and the high-frequency characteristics of the capacitor and the inductor are changed, and the filter cannot be normally filtered. At this time, the magnetic ring is used to suppress noise; the high-frequency radiation segment is 400 MHz–1 GHz, and the metal shielding sleeve is selected according to the influence of the shielding on the capacitive coupling. The topology of the power line integrated conducted/radiated EMI filter is shown in Fig. 3.36. In the process of rectification, the appropriate materials can be selected according to the correspond-

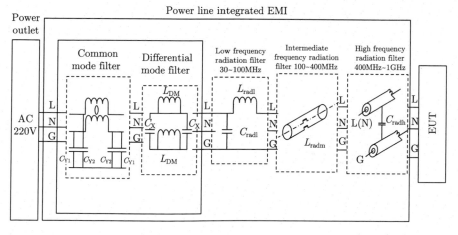

Fig. 3.36 Power line integrated conducted/radiated EMI filter

ing over-standard frequency bands in the test report, which provides a good reference for electromagnetic compatibility rectification.

3.3 Suppression Case Study of Conducted EMI Noise

3.3.1 Case #No. 1

1. Description of the Problem

Figure 3.37 shows a certain type of power concentrator produced by a company. The concentrator's function is to collect data from electric meters. First, collect this device through carrier or 485 communication. Then, transmit it to the main station by cable (optical fiber) or wireless network to improve the efficiency of meter reading and data integrity.

Because the power concentrator is widely used, if the conducted EMI noise exceeds the standard limit, it can seriously restrict the working characteristics of the equipment and affect the reading accuracy of the equipment. Therefore, the con-

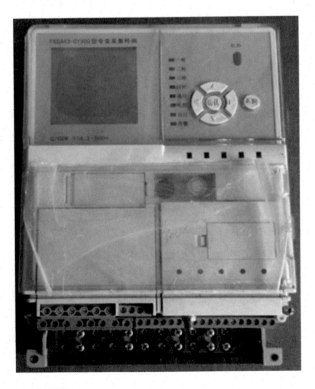

Fig. 3.37 Power concentrator

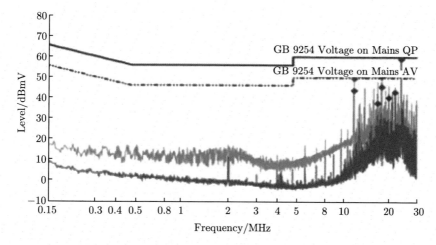

Fig. 3.38 Conducted EMI noise in the power concentrator before modification

ducted EMI noise needs to meet the GB 9254 Class B standard of our country. To test the conducted EMI noise characteristics, the R&S company's artificial power network (ENV216) and EMI receiver ESL3 (provided by the Jiangsu Electrical Equipment Electromagnetic Compatibility Engineering Laboratory) were used. The results of the test are shown in Fig. 3.38.

2. Measurement Standards

Electronic equipment in the range of information technology equipment sold and used in China's market needs to meet the Chinese national standard GB 9254—information technology equipment–radio EMI noise characteristics—limits and methods of measurement. The conditions of information technology equipment are as follows: ① The main function is to be able to record, store, display, retrieve, transfer, process, exchange, or control (or a combination of several functions) data and telecommunication messages. The device can be configured with one or more terminal ports that are typically used for messaging; ② Rated voltage does not exceed 600 V. According to the condition, the information technology equipment includes data processing equipment, office equipment, electronic commercial equipment, telecommunication equipment, etc. Obviously, the power concentrator is a conventional information technology equipment, and the conducted emission needs to meet the requirements of the GB 9254 standard and must meet the Class B standard.

3. Measurement Environment and Conditions

The conducted emission test of this case is the measurement equipment of the Electromagnetic Compatibility Engineering Laboratory of Jiangsu Electric Equipment. The measuring device uses an artificial power supply network (ENV216) manufactured by Rohde & Schwarz, Germany. The measuring receiver uses the ESL3 manufactured by R&S, Germany, and the test band is 9 kHz–2.7 GHz, with 0.5 dB level accuracy, 1 dB compression point +5 dB m, RF input anti-pulse power up to 10 mW, average noise display level (preamplifier on) <-152 dB m (1 Hz), resolution 10 Hz to 10 MHz $(-3$ dB); 200 Hz, 9 kHz, 120 kHz $(-6$ dB), 1 MHz (pulse bandwidth).

According to the standard requirements, the test site should be able to distinguish between EMI noise and environmental noise from the EUT. Site suitability in this regard can be determined by measuring the ambient noise level (EUT does not work) and should ensure that the noise level is at least 6 dB below the standard limit. When the combined result of both environmental noise and EMI noise does not exceed the specified limit, it is not necessary to require the ambient noise level to be 6 dB below the specified limit. In this case, the emission of the source can be considered to meet the requirements of the specified limits. When the combined result of both environmental noise and EMI noise exceeds the specified limit, it cannot be judged that the EUT does not meet the limit requirement, unless it indicates that the following two conditions are met at each frequency point corresponding to the limit: À The ambient noise level is at least 6 dB lower than the EMI noise plus the ambient noise level; Á The ambient noise level is at least 4.8 dB below the specified limit.

The noise level of the experimental site used in this experiment is lower than the standard limit of Class B by more than 6 dB, which meets the requirements of the test environment.

4. Conducted EMI Noise Suppression Analysis

According to the analysis of the test results in Fig. 3.38, the conducted EMI noises at the 12, 18, and 24 MHz frequencies were 51, 52, and 60 dB μV, which exceeds the GB 9254 Class B standard limit of 1, 2, and 10 dB μV, as shown in Table 3.2, and the maximum over-standard frequency is 24 MHz, which exceeds 10 dB μV.

The following problems are found:

(1) The parallel wires on the PCB flow through the high-frequency current signal, which causes crosstalk problems and generates conducted EMI noise, as shown in Fig. 3.39a. In free space, the RF electromagnetic field caused by the signal in the cable is

$$E_\theta = j \frac{I l Z_0 \beta_0 \sin \theta}{4\pi r} e^{-j\beta_0 r} \tag{3.30}$$

where Z_0 is the free space wave impedance, Ω; l is the cable length; I is the current in the cable; r is the test distance; β_0 is $2\pi/\lambda$, m^{-1}. As shown in Fig. 3.40,

Table 3.2 Electronic concentrator conducts EMI noise

Frequency (MHz)	Before suppression (dB μV)	Excess (dB μV)	After suppression (dB μV)	Safety margin (dB μV)
12	51	1	20	30
18	52	2	15	35
24	60	10	35	15

Fig. 3.39 PCB board level

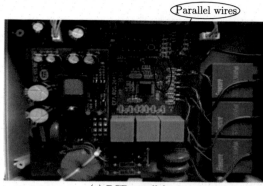

(a) PCB parallel wires

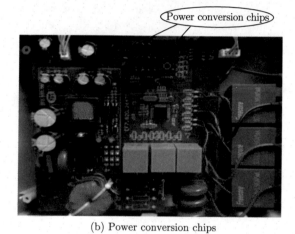

(b) Power conversion chips

Fig. 3.40 Cable crosstalk schematic

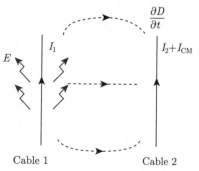

the radio frequency electromagnetic field E caused by the cable 1 is coupled to the cable 2 in the form of a spatial displacement current, thereby generating a common mode noise current I_{CM} in the cable 2.

Crosstalk is the conducted EMI noise caused by the coupling of RF electromagnetic fields generated by noise sources (including other cables, crystal oscillators or chips, etc.) near the cable into the cable. Therefore, the noise source of crosstalk is a space RF electromagnetic field. However, in the actual functional circuit, the working state of other cables, crystal oscillators or chips cannot be changed, and the electromagnetic shielding measures of the cable are also difficult to apply (low shielding effectiveness and high cost).

(2) The power conversion chip inside the power concentrator generates conducted EMI noise due to poor grounding. For common mode noise generated between the live/neutral line and ground, as shown in Fig. 3.41a, it can be equivalent to the common mode noise source U_{CM}, the common mode noise source internal impedance Z_{CM}, and the load impedance Z_{load}. The noise voltage U_{load} detected on the load is the conducted EMI noise.

$$U_{load} = \frac{Z_{load}}{Z_{CM} + Z_{load}} U_{CM} \tag{3.31}$$

When the PCB circuit is poorly grounded, there is a parasitic capacitance C between the floating ground and the ground. Therefore, the equivalent load Z_{load} is connected in series with the parasitic capacitance C, which is Z_L, as shown in Fig. 3.41b. Currently, the conducted EMI noise becomes U'_{load}.

$$U'_{load} = \frac{Z_L}{Z_{CM} + Z_L} U_{CM} = \frac{Z_{load} + \dfrac{1}{j\omega c}}{Z_{CM} + Z_{load} + \dfrac{1}{j\omega c}} U_{CM} \tag{3.32}$$

Under the condition that the common mode noise source U_{CM}, the common mode noise source internal impedance Z_{CM} and the load impedance Z_{load} are determined, U'_{load} is greater than U_{load}. U_{load} is the conducted EMI noise caused by the PCB circuit, and the increment ΔU_{load} is the conducted EMI noise caused by poor grounding.

$$\Delta U_{load} = U'_{load} - U_{load} \tag{3.33}$$

In practical applications, it is only necessary to strengthen the grounding system of the PCB circuit to effectively suppress conducted EMI noise caused by poor grounding. However, for complex electronic systems, there are some chips and their DC power supplies in the PCB circuit (the voltage levels are also different, such as 5, 3.3, 1.8, 1.2 V). To prevent mutual interference between the above DC power sources, it is necessary to improve the grounding system of the power supply to design a DC power supply between the chips.

Fig. 3.41 Conducted EMI noise caused by poor grounding

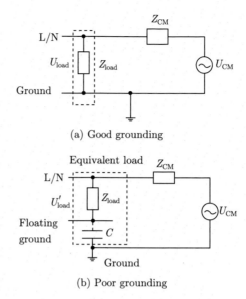

(a) Good grounding

(b) Poor grounding

5. Modification and Recommendations

Firstly, to improve the conducted EMI problem caused by the crosstalk problem, a crosstalk choke as shown in Fig. 3.42b is designed. For crosstalk, the cable is equivalent to a receiving antenna (short straight antenna, electric dipole). If the common mode choke coil is wound as shown in Fig. 3.42a, since the positions (field points) of the two antennas (the live line and the neutral line) are the same, and the amplitude and phase of the received RF electromagnetic field are the same, which greatly increases the influence of crosstalk. To solve the above problems, the crosstalk choke designed can effectively suppress the conducted EMI noise.

Then, in order to solve the problem of conducted EMI caused by poor grounding, a PCB DC power grounding system as shown in Fig. 3.43 was designed, where $U_1 = U_2, C_1 = C_2 = 22\,\mu\text{F}, C_3 = C_4 = C_5 = C_6 = 0.1\,\mu\text{F}, C_7 = 100\,\text{pF}, C_8 = 10\,\text{pF}$.

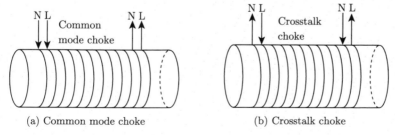

(a) Common mode choke

(b) Crosstalk choke

Fig. 3.42 Common mode choke and crosstalk choke

Fig. 3.43 PCB DC power grounding system

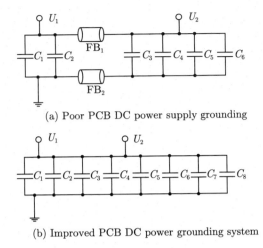

(a) Poor PCB DC power supply grounding

(b) Improved PCB DC power grounding system

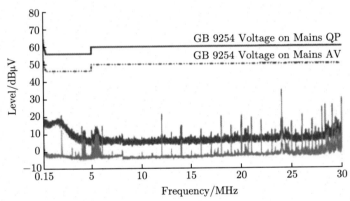

Fig. 3.44 Power concentrator conducted EMI noise after modification

Through the adoption of the above measures, the conducted EMI problem of the device has been effectively solved. The conducted EMI noise is significantly reduced, and the noise suppression effect can reach 37 dB μV, which conforms to the GB 9254 Class B standard, as shown in Fig. 3.44.

Recommendation: When designing the power supply, it is necessary to properly filter the external working power supply and the working power of the chip. Because it is easy to couple with high-frequency noise signals, it is transmitted throughout the PCB board. It can get greatly amplified once it passes through a large loop or connecting line.

3.3.2 Case #No. 2

1. Description of the Problem

Figure 3.45 shows a medical nebulizer, which is mainly used to treat respiratory diseases. The experiment was tested using the R&S artificial power network (ENV216) and the EMI receiver ESL3.

According to the GB 9254 standard, the tested frequency range is 150 kHz–30 MHz, the standard line in the 150 kHz–5 MHz band is 47 dB μV/m, and the standard line in the 5 MHz–30 MHz band is 50 dB μV/m. The original conducted EMI test results are shown in Fig. 3.46. The high-frequency noise exceeds the standard and does not pass the GB/T 6113.1—1995 test standard.

Fig. 3.45 Multi-function vaporizer

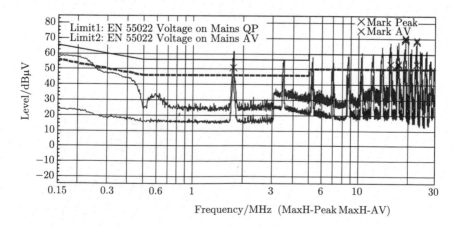

Fig. 3.46 Conducted EMI noise test results of the vaporizer

2. Inhibition Theory Analysis

It can be seen from the result that the over-standard frequency band is mainly in the frequency band of 1.7 MHz. Therefore, the problem is mainly due to the PCB. The characteristic impedance mismatch caused by the root line causes the noise signal in the impedance mismatch to be reflected and transmitted multiple times in the cable, causing higher harmonics of the main frequency of the signal to oscillate in the circuit. If the noise from the atomizer is suppressed, the noise on the octave is easily reduced. However, it is necessary to suppress the noise of 1.7 MHz below 47 dBμV/m without hindering the normal operation of the vaporizer. In addition, the 150–500 kHz band is also exceeded, and the noise in this band may mainly come from the power line.

3. Analysis of Conducted EMI Noise Suppression Methods

Aiming at the above analysis problem, the vaporizer sheet is the main noise source, and the root of the problem is the PCB cable impedance mismatch, which causes the main frequency signal and higher harmonics to oscillate in the circuit. The transformer is the necessary propagate for noise to reach LISN. According to the impedance matching principle, an impedance matched EMI filter is added to the primary and secondary transformer, which also suppresses the noise from the power supply. The primary side filter mainly consists of a common mode inductor and a safety capacitor. The ceramic capacitor is composed of a secondary side filter composed of a common mode inductor, a ceramic capacitor, and a magnetic ring. The results are shown in Fig. 3.47.

Through the above methods, the noise in the frequency range of 150–500 kHz is well suppressed. Although the noise on the 1.7MHz and the double frequency band is not suppressed below the standard line, the high-frequency noise on the high-frequency band is suppressed effectively.

In order to better suppress the noise in the 1.7 MHz band, more modifications need to be added on the path of the conducted noise propagation. Therefore, as shown in Fig. 3.48, there are chip clock signal pin, rectifier bridge, and switch on the PCB board. The device is connected to ceramic capacitors, such as 104, 102, and 101. Because the high-frequency operation of the switch generates a large number of

Fig. 3.47 Modification of the vaporizer

Fig. 3.48 After adding a magnetic ring

Fig. 3.49 Cable processing

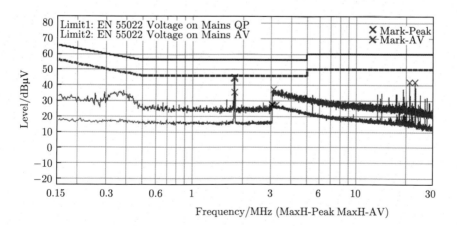

Fig. 3.50 Conducted EMI noise test results after modification

Table 3.3 Conducted EMI noise before and after modification

Excessive frequency point (MHz)	Noise suppression before (dB μV)	Super-scale amplitude (dB μV)	After noise suppression/dB μV)	Safety margin (dB μV)
1.806	74.62	11.25	42.15	21.22
18.306	77.01	16.92	43.68	16.40
19.410	75.92	18.27	43.55	14.10
19.514	73.86	17.86	44.08	11.92
20.610	72.56	16.56	44.05	11.95
21.718	68.34	12.34	42.02	13.98
23.814	62.27	6.27	39.95	16.05
24.918	67.03	11.03	42.36	13.64
25.022	66.98	10.98	39.24	16.76

harmonics, the capacitor can absorb some harmonics. As shown in Fig. 3.49, insert the magnetic ring on the power line and signal line and make sure the primary and secondary cables of the transformer must be kept at a certain distance. They are not allowed to be crossed and have to be on two levels. The EMI filter is separated. This is because the power line is too close or even crossed. The RF electromagnetic field is generated around the cable, so that the high-frequency noise is coupled to nearby lines. The interference between them is the source of crosstalk, that is, conducted EMI noise due to crosstalk.

4. Final Result After Modification

After the modification, the noise of the entire frequency band is reduced below the standard line, and the test passes. The result is shown in Fig. 3.50. The suppression result of the over-standard frequency point is shown in Table 3.3.

References

1. Nave MJ. Power line filter design for switched-mode power supplies. 2nd ed. Hoboken: Wiley; 1991.
2. Guo T, Chen DY, Lee FC. Separation of the common-mode and differential-mode conducted EMI noise. IEEE Trans Power Electron. 1996;11(3):480–8.
3. Paul CR, Hardin BK. Diagnosis and reduction of conducted noise emissions. IEEE Trans Electromagn Compat. 1988;30(4):553–60.
4. See KY. Network for conducted EMI diagnosis. IEEE Electron Lett. 1999;35(17):1446–7.
5. Mardiguian M, Raimbourg J. An alternate, complementary method for characterizing EMI filters. IEEE Int Symp Electromagn Compat. 1999;10(3):171–3.
6. Lo YK, Chiu HJ, Song TH. A software-based CM and DM measurement system for the conducted EMI. IEEE Trans Ind Electron. 2000;47(4):977–8.

7. Zhao Y, See KY, Li SJ. Noise diagnosis techniques in conducted electromagnetic interference (EMI) measurement: methods analysis and design. In: Antennas and propagation society international symposium. CA: Monterey; 2004.
8. Zhao Y, See KY. Performance study of CM/DM discrimination network for conducted EMI diagnosis. Chin J Electron. 2003;12(4):536–8.
9. Zhao Y, See KY. Noise signal mode discrimination network research for conducted EMI emission. J Nanjing Normal Univ Eng Ed. 2002;2(4):26–9.
10. Zhao Y. Discussion on conducted EMI noise signal mode separation and noise suppression. J Nanjing Normal Univ Eng Ed. 2004;4:1–4.
11. Li SJ. Intellectual measurement system study for power-line noise diagnosis and suppression. J Nanjing Normal Univ Eng Ed. 2005;2:5–9.

Chapter 4
Radiated EMI Noise Generation Mechanism, Measurement, and Diagnosis

4.1 Generation Mechanism and Analysis of Radiated EMI Noise

4.1.1 Common Mode and Differential Mode Definition

Radiated EMI noise from digital electronic devices occurs in differential and common modes. Differential mode radiated EMI noise is the result of the normal operation of the circuit, which is generated by the current flowing through the loop formed by the circuit conductor. These loops act as the main radiated magnetic field of the small ring antenna. Although these signal loops are necessary for circuit work, their size and area must be controlled in the design process in order to minimize radiated EMI noise.

However, common mode radiated EMI noise is a parasitic phenomenon of the circuit, which is caused by unwanted voltage drop on the conductor. The differential mode current flowing through the grounding impedance produces a voltage drop in the digital logic grounding system. Therefore, when the cable is connected to the system, it is driven by common mode ground potential to form an antenna, and the radiated EMI noise is mainly electric field. Because these parasitic impedances are not intentionally designed into the system or displayed in the file, the common mode radiated EMI noise is often more difficult to understand and control. In the design process, the method of dealing with the common mode emission problem must be implemented.

1. **Differential Mode Radiated EMI Noise**

Differential mode radiated EMI noise is simulated as a small ring antenna. For a small loop with an area of A, which carries current I_{DM}, at a distance of r in the far field, the amplitude of electric field E measured in the free space is [1]

$$E = 131.6 \times 10^{-16} \left(f^2 A I_{DM} \right) \left(\frac{1}{r} \right) \sin \theta \tag{4.1}$$

© Science Press 2021
Y. Zhao et al., *Electromagnetic Compatibility*,
https://doi.org/10.1007/978-981-16-6452-6_4

Fig. 4.1 Radiation pattern
of a small ring antenna in
free space

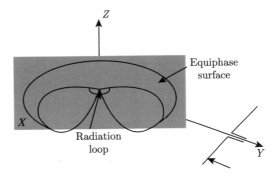

The unit of E is V/m, the unit of f is Hz, the unit of A is m [2], the unit of I_{DM} is A, unit of r is m, and θ is the angle between the observation point and the vertical line of the loop plane.

In a small ring with a circumference less than a quarter of the wavelength, the phase of the current is the same everywhere. For larger rings, currents are no longer in phase everywhere, so some currents may decrease rather than increase for the total emission.

As shown in Fig. 4.1, the pattern of a small ring antenna in free space is in a ring shape (or doughnut shape). The maximum radiated EMI noise comes from the edge of the ring and appears on the plane of the ring. Zero radiated EMI noise occurs in the normal direction of the annular plane. As shown in Fig. 4.1, because the electric field is polarized on the annular plane, the maximum electric field can be detected by a receiving antenna with the same polarization direction.

When the circumference of the ring increases by more than a quarter of the wavelength, the pattern shown in Fig. 4.1 is no longer applicable. For a ring whose circumference equals a wavelength, the radiation pattern rotates 90° so that the maximum radiated EMI noise appears in the normal direction of the ring plane. Therefore, the zero radiation direction of the small ring becomes the maximum radiation direction of the large ring.

Although Eq. (4.1) is derived from a circular ring, it can be applied to any planar ring because the radiation amplitude and pattern of the small ring are independent of the shape of the ring and only depend on the area of the ring. Whatever the shape of the ring is, all small rings with the same area radiate the same.

The first term in Eq. (4.1) is a constant describing the properties of the medium—the free space in the formula. The second term defines the characteristics of the radiation source, that is, the loop. The third term represents the delay of field propagation from the source. The last item describes the angle direction of the measuring point relative to the vertical line of the annular plane, that is, the angle deviating from the Z-axis in Fig. 4.1.

Equation (4.1) refers to the case where a small ring is located in free space, and there is no reflecting surface around it. However, most EMC measurements of radiation from electronic products are made on open ground, not in free space. The ground provides a reflection surface that must be considered. This reflecting surface can increase the measured emission by as much as 6 dB (or twice). Considering this reflection, Eq. (4.1) must be multiplied by coefficient 2. Correct ground reflection, and assume that observation ($\theta = 90°$) is made at r from the annular plane. Measurements of Eq. (4.1) on open sites can be written as [1, 2]

$$E = 263 \times 10^{-16} \left(f^2 A I_{DM}\right) \left(\frac{1}{r}\right) \tag{4.2}$$

Equation (4.2) shows that radiation is proportional to the square of current I, ring area A, and square of frequency f. Square of frequency provides career insurance for future EMC engineers.

For the test distance of 3 m, Eq. (4.2) can be written as [2, 3]

$$E = 87.6 \times 10^{-16} \left(f^2 A I_{DM}\right) \tag{4.3}$$

Therefore, differential mode (loop) radiated EMI noise can be controlled by the following methods:

(1) Reducing the current amplitude;
(2) Reducing the harmonic component of frequency or current;
(3) Reducing the area of ring road.

2. Common Mode Radiated EMI Noise

Differential mode radiated EMI noise can be controlled by PCB design and wiring. However, common mode radiated EMI noise is more difficult to control and usually determines the overall emission characteristics of a product.

The most common form of common mode radiated EMI noise is emitted by the cable of the system. Radiation frequency is determined by common mode voltage (usually ground voltage). In the case of common mode radiated EMI noise, what the cable does is not important. What matters is that it connects to the system and affects the ground of the system in some way. The radiation frequency is independent of the original signal in the cable.

Common mode emission can be simulated as a dipole or monopole antenna (cable) driven by a noise voltage (ground voltage). For a short dipole antenna of length l, the magnitude of the measured electromagnetic intensity in the far field with distance r from the source is as follows [3, 4]

$$E = \frac{4\pi \times 10^{-7} \left(f l I_{CM}\right) \sin \theta}{r} \tag{4.4}$$

The unit of E is V/m, the unit of f is Hz, I is common mode current in cable (antenna) with each unit of A, the unit of l and r is m, and θ is the angle between observation point and antenna axis. The maximum field intensity appears in the direction perpendicular to the axis of the antenna, namely $\theta = 90°$.

Fig. 4.2 Capacitor-loaded
antenna

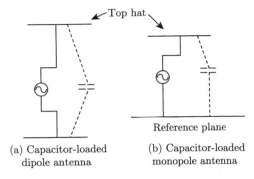

(a) Capacitor-loaded   (b) Capacitor-loaded
dipole antenna   monopole antenna

When the dipole antenna is placed along the "Z-axis", the free space pattern of a small dipole antenna is the same as that of the compact antenna in Fig. 4.1. For a monopole antenna on an infinite reference plane, the pattern and amplitude are the same as those of a dipole antenna only in the upper half space.

This makes Eq. (4.4) suitable for an ideal dipole antenna with uniform current distribution. For a real dipole antenna, the current at the open end of the line is zero. For a small antenna, the current distribution is linear along the length of the antenna. Therefore, the average current on the antenna is only half of the maximum current.

In fact, if a metal cap is installed at the open end of a dipole or monopole antenna, a more uniform current distribution can be obtained, as shown in Fig. 4.2. This increases the capacitance of the terminal, attracts more current to the terminal, and generates an almost uniform current distribution along the antenna. This kind of antenna is called capacitive loading or cap antenna. This structure is approximated when the antenna (cable) is connected to other components of the device. The cap antenna is very similar to the ideal uniform current antenna model, and Eq. (4.4) can be used.

For the case observed at a distance r in the direction ($\theta = 90°$) perpendicular to the antenna axis, using MKS units, Eq. (4.4) can be written as follows [4, 5]

$$E = \frac{12.6 \times 10^{-7} \, (f l I_{CM})}{r} \tag{4.5}$$

Equation (4.5) shows that radiation is proportional to frequency, antenna length, and the amplitude of common mode current on the antenna. The main way to reduce this radiation is to limit common mode current, which is not needed for the normal operation of the circuit.

Therefore, common mode (dipole) radiated EMI noise can be controlled by the following methods:

(1) Reducing the amplitude of common mode current;
(2) Reducing the harmonic component of frequency or current;
(3) Reducing the length of antenna (cable).

Fig. 4.3 Common mode
emission envelope with
frequency

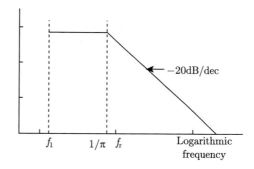

-20dB/dec

f_1 $1/\pi$ f_r Logarithmic
frequency

Table 4.1 Maximum common mode current permissible in a 50 MHz and 1-m-long cable

Standard	Limit value/(μV/m)	Distance/m	Maximum common mode current/μA
FCC class A	90	10	15
FCC B class	100	3	5
MIL-STD 461	16	1	0.25

For current waveforms different from sinusoidal waves, the Fourier series of current must be determined before substitution (4.5).

The frequency term in Eq. (4.5) indicates that the frequency increases at a rate of 20 dB/dec. Therefore, based on the Fourier coefficients and Eq. (4.5), it reveals that the envelope of the common mode emission spectrum is flat before frequency of $1/\pi t_r$ and decreases at a rate of 20 dB/dec when the frequency is higher than that.

Figure 4.3 shows how the envelope of common mode emission varies with frequency. We can think that common mode emission is more like a low-frequency problem, while differential mode emission is more like a high-frequency problem. For the rise time in the range of 1–10 ns, most common mode emission occurs in the frequency band of 30–300 MHz.

Equation (4.5), I is [5]

$$I = \frac{0.8 E_{rm\,r}}{fl} \tag{4.6}$$

Among them, E_r is the electric field strength, the unit is μV/m, the unit of I is μA, the unit of f is MHz, and the unit of r and l is m.

Table 4.1 lists the approximate maximum common mode currents allowed in a 50 MHz and 1-m-long cable that do not exceed the specified emission regulatory limits.

The ratio of differential mode current to common mode current needed to generate the same radiation emission can be determined by Eq. (4.2) equaling Eq. (4.5), and the current ratio can be solved. Therefore [5, 6]

$$\frac{I_{DM}}{I_{CM}} = \frac{48 \times 10^6 l}{f A} \qquad (4.7a)$$

And I_{DM} and I_{CM} are the differential and common mode currents needed to generate the same radiation emission, respectively. If the length of the cable is equal to 1 m, the area of the ring road is equal to 10 cm^2 (0.001 m^2) and the frequency is 48 MHz, then Eq. (4.7a) is simplified to [5]

$$\frac{I_{DM}}{I_{CM}} = 1000 \qquad (4.7b)$$

Equation (4.7b) shows that the differential mode current is three orders of magnitude larger than the common mode current in order to generate the same radiated field. In other words, the common mode radiated EMI noise mechanism is much more effective than the differential mode radiated EMI noise mechanism. The common mode current of several microamperes can produce the same amount of radiation as the differential mode current of several milliamperes.

Equations (4.4) and (4.5) are derived from short cables. For long cables ($l > \lambda/4$), Eq. (4.5) can be written as [5, 7]

$$E = \frac{94.5 I_{CM}}{r} \qquad (4.8)$$

Among them, I_{CM} is the common mode current, r is the measuring distance, and the unit is m. It is noted that for long cables, the envelope of common mode radiation emission is not a function of cable length or frequency, but only of common mode current in cables.

4.1.2 Radiated EMI Noise Generation Mechanism

The high-frequency noise generated by electronic equipment is transmitted in space through radiation antenna. According to the transmission law of radiated electromagnetic field, it can be divided into near-field area and far-field area.

If the wavelength of the radiated noise is λ, the near-field region is when the distance between the field point and the test site is less than $\lambda/2\pi$, and the far-field region is when the distance between the field point and the test site is greater than $\lambda/2\pi$, as shown in Fig. 4.4.

In the near-field region, the radiated electric field and the radiated magnetic field decrease with the increase of the test distance, and the radiated electric field and the radiated magnetic field have a nonlinear relationship; in the far-field region, the radiated electric field and the radiated magnetic field also decrease with the increase of the test distance, but there is a linear relationship between them, that is, the radiated electric field and the radiated magnetic field have a one-to-one correspondence.

Fig. 4.4 Near- and far-field radiation

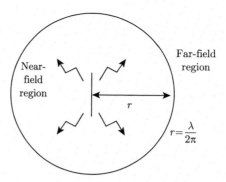

Therefore, in the process of EMI noise measurement, it is very important to select the appropriate test distance.

However, GB 9254 and other standards stipulate that the test points of radiated EMI noise should be in the far-field area, such as open field of 10, 5, and 3 m. Because the lower limit of EMI noise test band is 30 MHz, the corresponding wavelength is 10 m, and the distance between the near field and the far field is [7]

$$r_0 = \frac{\lambda}{2\pi} = \frac{5}{\pi} \approx 1.59 \tag{4.9}$$

In the formula, r_0 is the decomposition distance of the near field and the far field, the unit is m. Equation (4.9) shows that the existing 10, 5, and 3 m anechoic chambers can meet the requirements of far-field testing for test distance.

Radiated EMI noise is generated by high-speed digital circuits and radiation antennas. Different radiation antennas correspond to different electromagnetic propagation characteristics. In the far-field region, the radiated electric field and radiated magnetic field have a linear relationship, so it cannot reflect the corresponding electromagnetic propagation characteristics of different radiation antennas. Namely only the radiated field intensity can be obtained by far-field testing according to the standard, but the radiation characteristics cannot be revealed.

In general, if the EUT is located at one-dimensional coordinate x', the test point is located at x, the length of the EUT radiation antenna is l, and the frequency corresponding to the radiated EMI noise caused by EUT is f, and the distance between the radiation antenna and the test site can be expressed as follows [8]

$$r \approx R - \boldsymbol{n} \cdot x' \tag{4.10}$$

In the formula, R is the distance between the coordinate origin and the test site, and $\boldsymbol{n}$ is the unit vector in the direction of R.

The corresponding wavelength of radiated EMI noise is [7, 8]

$$\lambda = \frac{c}{f} \tag{4.11}$$

In the formula, c is the speed of light in vacuum, and λ is the wavelength corresponding to radiated EMI noise.

$$l \ll \lambda, \quad l \ll r \tag{4.12}$$

If Eq. (4.12) is satisfied, the above radiation is of small size. In general, if the current density in the radiation antenna is J, then [9, 10]

$$A(x, t) = \frac{\mu_0}{4\pi} \int \frac{J\left(x', t - \frac{r}{c}\right)}{r} du' \tag{4.13}$$

In Eq. (4.13), A is the postponement potential. The current density J can be separated from the component.

$$J(x', t) = J(x') \text{ rm } e^{-j\omega t} \tag{4.14}$$

By substituting Eq. (4.14) into Eq. (4.13), there is [10]

$$A(x, t) = \frac{\mu_0}{4\pi} \int \frac{J(x') e^{j(kr - \omega t)}}{r} du' \tag{4.15}$$

In the formula, k is the wave vector. Separating component of delayed potential A can obtain [10]:

$$A(x, t) = A(x) e^{-j\omega t} \tag{4.16}$$

Substituting Eq. (4.16) into Eq. (4.15):

$$A(x) = \frac{\mu_0}{4\pi} \int \frac{J(x') e^{jkr}}{r} du' \tag{4.17}$$

Substituting Eq. (4.10) into Eq. (4.17):

$$A(x) = \frac{\mu_0}{4\pi} \int \frac{J(x') e^{jk(R - n \cdot x')}}{R - n \cdot x'} du' \tag{4.18}$$

Item $-n \cdot x'$ in denominator of Eq. (4.18) can be omitted and $e^{-jkn \cdot x'}$ expanded according to $kn \cdot x'$.

$$A(x) = \frac{\mu_0 e^{jkR}}{4\pi R} \int J(x') \left(1 - jkn \cdot x' + \cdots\right) du' \tag{4.19}$$

The first and second terms in Eq. (4.19) can be expressed as common mode radiation of dipole and differential mode radiation of magnetic dipole, respectively.

Fig. 4.5 Radiated EMI
noise model of common
mode electric dipole

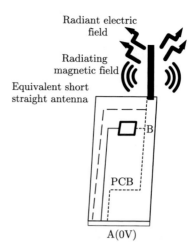

Therefore, according to the different electromagnetic propagation characteristics of the radiation antenna, the radiated EMI noise can be divided into the common mode radiated EMI noise of the electric dipole and the differential mode radiated EMI noise of the magnetic dipole. It is noteworthy that the common mode and differential mode radiation characteristics mentioned above are only reflected in the near-field region.

As shown in Fig. 4.5, for common mode radiation of electric dipoles, the delay potential only needs the first term in the reserved Eq. (4.19) [10].

$$A\left(x\right) = \frac{\mu_0 e^{jkR}}{4\pi R} \int J\left(x'\right) \mathrm{d}u' \tag{4.20}$$

If there are n_i particles with v_i velocity per unit volume in the radiation antenna, and the charged quantity is q_i particles, then

$$J = \sum_i n_i q_i v_i \tag{4.21}$$

It can be obtained from Eq. (4.20) and Eq. (4.21) that [10, 11]

$$\int J\left(x'\right) \mathrm{d}V' = \sum_i q_i v_i \tag{4.22}$$

Because

$$\dot{p} = \sum_i q_i v_i \tag{4.23}$$

In the formula, $\dot{p}$ is the first derivative of dipole moment to time, Eq. (4.23) is substituted for Eq. (4.22).

Fig. 4.6 Radiated EMI
noise model of differential
mode magnetic dipole

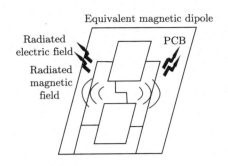

$$\int \boldsymbol{J}\left(x'\right)\mathrm{d}V' = \dot{p} \tag{4.24}$$

It can be obtained from Eqs. (4.21) and (4.24).

$$A\left(x\right) = \frac{\mu_0 \mathrm{e}^{\mathrm{j}kR}}{4\pi R}\dot{p} \tag{4.25}$$

The magnetic field intensity, electric field intensity, energy flux density, and radiation power of common mode radiation of electric dipole can be obtained from Eq. (4.25) [10].

$$B_{CM} = \frac{1}{4\pi\varepsilon_0 c^3 R}\,|\ddot{p}|\,\mathrm{e}^{\mathrm{j}kR}\sin\theta e_\phi, \quad E_{CM} = \frac{1}{4\pi\varepsilon_0 c^2 R}\,|\ddot{p}|\,\mathrm{e}^{\mathrm{j}kR}\sin\theta e_\theta$$

$$S_{CM} = \frac{1}{32\pi^2\varepsilon_0 c^3 R^2}\,|\ddot{p}|^2\sin^2\theta\boldsymbol{n}, \quad P_{CM} = \frac{1}{4\pi\varepsilon_0}\frac{|\ddot{p}|^2}{3c^3} \tag{4.26}$$

From Eq. (4.26) and Maxwell's formulae, taking the direction of the maximum radiated field gradient as the test direction, the common mode radiated EMI noise generated by the electric dipole at the far field r is as follows:

$$E_{CM} = 12.6 \times 10^{-7}\frac{f l I_{CM}}{r} \tag{4.27}$$

In the formula, I_{CM} is the common mode current in the radiation antenna, l is the length of the radiation antenna, and r is the test distance.

Therefore, radiated EMI noise can be described by Eqs. (4.26), (4.27) and Fig. 4.5.

As shown in Fig. 4.6, for the differential mode radiation of a magnetic dipole, the delay potential only needs the second term in the retention Eq. (4.19).

$$A\left(x\right) = \frac{-\mathrm{j}k\mu_0 \mathrm{e}^{\mathrm{j}kR}}{4\pi R}\int \boldsymbol{J}\left(x'\right)\left(\boldsymbol{n}\cdot x'\right)\mathrm{d}u' \tag{4.28}$$

Considering the electric and magnetic dipole moments, the integrand function of Eq. (4.28)

$$J'\left(n \cdot x'\right) = n \cdot x' J' \tag{4.29}$$

In the above formula, $x' J'$ represents tensors [10]

$$x' J' = \frac{1}{2}\left(x' J' + J' x'\right) + \frac{1}{2}\left(x' J' - J' x'\right) \tag{4.30}$$

In the formula, $(x' J' + J' x')/2$ is the symmetric part of the tensor, and $(x' J' - J' x')/2$ is the anti-symmetric part of a tensor. If Eq. (4.30) is substituted into Eq. (4.28), then

$$\int J(x')(n \cdot x') du' = \frac{1}{2}\int \left[(n \cdot x') J' + (n \cdot J') x'\right] du' + \frac{1}{2}\int \left[(n \cdot x') J' - (n \cdot J') x'\right] du' \tag{4.31}$$

Consider the second term in Eq. (4.31) to get:

$$\left(n \cdot x'\right) J' - \left(n \cdot J'\right) x' = -n \times \left(x' \times J'\right) \tag{4.32}$$

The second term in Eq. (4.31) is

$$-n \times \int \frac{1}{2}\left(x' \times J'\right) du' = -n \times m \tag{4.33}$$

In the formula, m is the magnetic dipole moment, while the first term in Eq. (4.30) is the electric dipole moment, which is usually negligible.

$$A\left(x\right) = \frac{jk\mu_0 e^{jkR}}{4\pi R}\left(n \times m\right) \tag{4.34}$$

The magnetic field intensity, electric field intensity, energy flux density, and radiation power of magnetic dipole differential mode radiation can be obtained from Eq. (4.34).

$$B_{DM} = \nabla \times A = \frac{\mu_0 e^{jkR}}{4\pi c^2 R}\left(\ddot{m} \times n\right) \times n \quad E_{DM} = cB \times A = -\frac{\mu_0 e^{jkR}}{4\pi c R}\ddot{m} \times n$$

$$S_{DM} = \frac{\mu_0 \omega^4 |m|^2}{32\pi^2 c^3 R^2}\sin^2 \theta n \qquad P_{DM} = \frac{\mu_0 \omega^4 |m|^2}{12\pi c^3} \tag{4.35}$$

From Eq. (4.35) and Maxwell's formulae, considering the reflection of grounding copyright, the radiated EMI noise generated by the magnetic dipole at the far field r is as follows [10]:

$$E_{DM} = 2.632 \times 10^{-14}\frac{f^2 A I_{DM}}{r} \tag{4.36}$$

In the formula, I_{DM} is the differential mode current in the radiation antenna, A is the area of the radiation antenna, and r is the test distance.

Therefore, magnetic dipole differential mode radiated EMI noise can be described by Eqs. (4.35), (4.36) and Fig. 4.6.

4.1.3   Equivalent Circuit of Radiated EMI Noise

Because the performance of electromagnetic compatibility directly restricts the efficiency of electronic and electrical equipment and even leads to electrical accidents. The problem of radiated electromagnetic interference is one of the main problems. However, because the structure of electronic equipment is generally complex, and the processing speed of integrated circuit chips is faster and faster, even more than 100 MHz, the mechanism of EMI noise is more complex, which includes common mode radiated interference (such as high-speed digital processing chip ARM9, active or passive crystal oscillator, cable) and differential mode radiated interference (such as signal loop, power loop). It also includes common mode/differential mode hybrid interference (such as structural interference). How to diagnose the radiated electromagnetic interference mechanism of electronic equipment reasonably is related to how to solve the problem of radiated electromagnetic interference quickly and efficiently. The key to diagnose the radiated electromagnetic interference mechanism is to model the characteristics of the radiated electromagnetic interference noise.

High-frequency common mode radiated noise signals are generated by high-speed processing chip and clock crystal oscillator in the working process. Figure 4.7a shows the interference model of conventional high-speed processing chip. Port ① is the clock output port of the chip, and the signal output is transmitted through the connecting cable as one of the main noise interference sources. According to the radiation mechanism, the equivalent interference circuit model as shown in Fig. 4.7b can be obtained. The $U_{N(CLK)}$ is the data transmission signal of port ①, I_{CM} is the current signal corresponding to the signal on the transmission line, and Z_{CM} is the equivalent common mode impedance. Then

$$I_{CM} = \frac{U_{N(CLK)}}{Z_{CM}} \tag{4.37}$$

Because port ① is a clock signal and rectangular wave signal in theory, but because of the complex multi-band characteristics of signal data processed by high-speed signal processing, the clock signal in this port is actually composed of a variety of frequency signals. The largest interference signal is the main frequency U_0 of the clock signal, and the others are mixed interference signals $U_i (i = 1, 2, \ldots, k)$. Namely:

$$U_{N(CLK)} = U_0 + \sum_{i=1}^{k} U_i, i = 1, 2, 3, \ldots, k \tag{4.38}$$

Fig. 4.7 Chip interference model

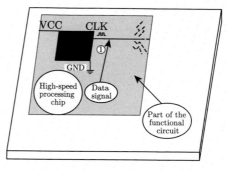

(a) Interference structure diagram of PCB chips

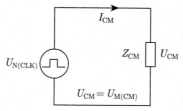

(b) Common mode equivalent interference model circuit

For the main frequency clock signal, assuming the rising edge is t_{r1}, then BW $= 1/\pi t_{r1}$, and for k mixed interference signals, assuming the rising edge is t_{ri}, then BW $= 1/\pi t_{ri}$ $(i = 1, 2, 3, \ldots, k)$.

Once the noise interference signal $U_{N(CLK)}$ is transmitted through the connecting cables between chips, it can be amplified to a certain extent. Combined Eqs. (4.37) and (4.8), the spatial radiated field strength can be obtained as follows [10, 11]:

$$
\begin{aligned}
E &\approx \frac{2U_{CM}}{r}\sqrt{\frac{30}{Z_{CM}}} \\
&\approx \frac{2U_{N(CLK)}}{r}\sqrt{\frac{30I_{CM}}{U_{N(CLK)}}} \\
&\approx \frac{2}{r}\sqrt{30U_{N(CLK)}I_{CM}} \\
&= \frac{2}{r}\sqrt{30\left(U_0 + \sum_{i=1}^{k}U_i\right)I_{CM}U_0}, \quad i = 1, 2, 3, \ldots, k
\end{aligned}
\tag{4.39}
$$

where r is the test distance.

Figure 4.8a shows the physical figure of 24.56 MHz active crystal oscillator. The four pins are power supply, ground, clock output, and vacant port, respectively. The biggest difference between it and the conventional passive crystal oscillator is that DC

Fig. 4.8 Radiation model of
the chip

(a) Active crystal oscillator

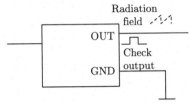

(b) Equivalent interference model circuit

power supply is needed and the supply voltage varies from 6 to 26 V. When the signal of the output port of the signal is transmitted to the corresponding processing chip through the connecting cable for normal working period, transmission on the cable is one of the most important interference sources for radiated EMI noise. The main frequency clock signal is transmitted in the equipment with the normal operation of the system. Once the signal is transmitted on the long-distance cable, it also produces a larger electromagnetic field strength in space. According to the working principle, the equivalent interference model circuit as shown in Fig. 4.8b can be obtained. The radiated EMI characteristics are like Eq. 4.39 derived from the chip interference model.

$$E \approx \frac{2U_{\text{out}}}{r} \sqrt{\frac{30 I_0}{U_{\text{out}}}} \approx \frac{2}{r} \sqrt{30 U_{\text{out}} I_0} \qquad (4.40)$$

The function chips and circuit modules on PCB board depend on useful signals to realize board-level functions, but these useful signals are often mixed with many high-frequency useless signals in the transmission process. After amplification (or attenuation) of these useless signals by different functional modules, noise is amplified on PCB board-level transmission lines, as shown in Fig. 4.9a as board-level interference PCB model. According to the structure model, the equivalent circuit interference model can be obtained, as shown in Fig. 4.9b. The original noise signal is U_0, which passes through modules 1, 2, 3, … transmission and transformation of N, and the final noise signal at the output end can be assumed as follows:

$$U_{\text{N}} = A_1 \cdot A_2 \cdot A_3 \cdots \cdot A_n U_0 \qquad (4.41)$$

Among them $A_1, A_2, A_3, \ldots, A_n$ are the signal transmission coefficients of each module which the noise interference signal passes, and the noise signal at the output port is dominated by common mode interference voltage.

$$E \approx \frac{2A_1 A_2 A_3 \cdots\cdots A_n U_0}{r} \sqrt{\frac{30}{Z_{CM}}} \tag{4.42}$$

According to the equivalent circuit shown in Fig. 4.9b, from the point of view of circuit analysis, the output signal of A_n module seems not directly related to the interference source signal U_0, but without the contribution of U_0, the interference signal cannot be contained at the output port of A_n module at all, so it does not radiate to the space field through cable antenna, which leads to the radiation exceeding the standard of the equipment. In order to solve this problem, it is necessary to diagnose the noise source first, that is, to determine the interference source through the voltage probe and then to diagnose and analyze the A_n module with the voltage probe to confirm whether the output contains the U_0 noise signal or the frequency doubling interference. There are usually two kinds of restraint measures:

(1) Radiated EMI filtering (beads, capacitors, etc.) is used in U_0 output.
(2) Radiated EMI filtering is used to filter the output of A_n.

Cable radiation is one of the most fundamental problems in the radiation exceeding standard of electronic equipment. Most of the radiation exceeding standard of electronic equipment is caused by the connecting cables (including data cables, power cables, etc.) inside the system. Even if the high-frequency signal cannot be transmitted in the PCB functional circuit, if the quality of the chip passes, there are no connecting cables or shorter connecting cables (including PCB) inside the electronic system. In the case of the above wiring on the board, the radiation problem does not exceed the standard or be easier to solve. Cable interference noise mainly consists of connecting cables and power cables, and the radiation model of such cables can be divided into two types: common mode radiated interference and differential mode radiated interference. Common mode noise is formed without a loop. Once a loop is formed, there is differential mode interference noise. Differential mode interference is shown in Fig. 4.10. According to the equivalent interference model, the spatial radiated field strength can be obtained [10].

$$E \approx \frac{2U_{DM}}{r} \sqrt{\frac{30}{Z_{DM}}} \tag{4.43}$$

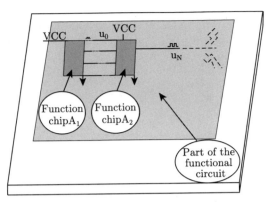

(a) PCB structure model of multistage chips

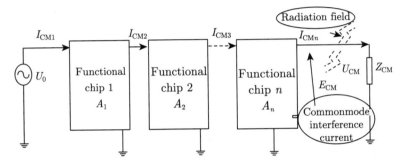

(b) Equivalent interference model circuit

Fig. 4.9 PCB board-level interference model

Fig. 4.10 Differential mode radiation equivalent circuit

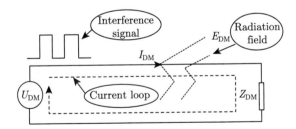

4.2 Radiated EMI Noise Measurement Method

4.2.1 Anechoic Chamber Classification and Working Principles

The anechoic chamber is primarily used to simulate open fields and is used for sealed shielded rooms that radiate radio interference (EMI) and electromagnetic sensitivity

(EMS) measurements. The anechoic chamber is like a steel fully enclosed house, including six-sided shells, doors, windows, and other general housing elements. It requires strict electromagnetic sealing performance. The cold-rolled steel sheet is the main shielding material, and all the inlet and outlet pipelines are shielded accordingly in order to block electromagnetic radiation in and out.

The working principle of the anechoic chamber is to use the ferrite absorbing material to absorb electromagnetic waves to eliminate electromagnetic interference in the environment.

According to the function of the design of the darkroom, the test site is divided into three types: all-wave darkroom, semi-anechoic darkroom, and open field. Radiation tests conducted in these three test sites are generally considered to be consistent with the propagation of electromagnetic waves in free space.

A full-wave anechoic chamber is shown in Fig. 4.11a. The full anechoic chamber is a shielded six-sided box with electromagnetic waves absorbing materials on the inner floor, walls, and ceiling. The absorbing material is generally a cone made of polyurethane foam. It absorbs the incident electromagnetic wave energy and greatly attenuates the scattering. However, due to the low-frequency performance of the cone absorbing material, ferrite and cone absorbing materials are usually used simultaneously to improve the low-frequency performance. And the full anechoic chamber reduces the interference of the external electromagnetic wave signal to the test signal, and the electromagnetic wave absorbing materials can reduce the multi-path effect caused by the reflection of the wall and the ceiling on the test result and are suitable for the experiment of emission, sensitivity, and immunity.

In practical use, if the shielding effectiveness of the shield can reach 80–140 dB, the interference to the external environment can be neglected, and the free space can be simulated in the full anechoic chamber. Compared to the other two test sites, the full anechoic chamber has a minimal floor, ceiling and wall reflections, and minimal interference from the outside environment, and is unaffected by outside weather. The disadvantage is that it is limited by cost and has limited test space.

A semi-anechoic chamber is shown in Fig. 4.11b. The semi-anechoic chamber is similar to the full anechoic chamber. It is also a shielded six-sided box covered with electromagnetic wave absorbing materials. The difference is that the semi-anechoic chamber uses conductive flooring and does not cover the absorbing materials. The semi-anechoic chamber simulates an ideal open field situation where the field has an infinitely large conductive ground plane. In a semi-anechoic chamber, since the ground does not cover the absorbing materials, a reflection path is generated such that the signal received by the receiving antenna is the sum of the direct path and the reflected path signal.

The open field is shown in Fig. 4.11c. The open area test (OAT) is an elliptical or circular test site with flat, open, uniform conductivity, and no reflection. The ideal open field has good conductivity, and the area is infinite at 30 MHz. The signal received by the receiving antenna between 1000 MHz is the sum of the direct path and the reflected path signal. However, in practical applications, although good ground conductivity can be obtained, the area of the open field is limited, and thus the phase difference between the transmitting antenna and the receiving antenna may be

(a) Full-wave anechoic chamber   (b) Semi-anechoic chamber

(c) Open field

Fig. 4.11 Three types of anechoic camber

caused. In the launch test, the open field is used in the same way as the semi-anechoic chamber.

Since the open field can only measure the radiation emission and cannot measure the radiation sensitivity, the electromagnetic environment in space is getting worse and worse, and it is difficult to find a pure open field. At present, electromagnetic compatibility rarely uses open test sites at home and abroad. A half-wave darkroom is typically used for testing.

4.2.2 Radiated EMI Noise Measurement Equipment and Composition

1. Measure Equipment and Requirements

In order to study radiated EMI, it is necessary to measure it. There are a variety of measurement equipment and measurement methods. The following is a brief introduction.

(1) Electric field radiation: receiver (below 1G), spectrum analyzer (1G or more), anechoic chamber, antenna (combination of double cone and logarithmic cycle below 1G or wide-band composite antenna, 1G, or more horn antenna);

(2) Magnetic field radiation: receiver, three-ring antenna, or single small loop antenna;

(3) Interference power: receiver, power absorbing clamp.

Among them, the measurement receiver quasi-peak measurement receiver below 1 GHz should meet the requirements of GB/T 6113.101—2016-Chapter 4. The peak measurement receiver shall comply with the requirements of GB/T 6113.101—2016-Chapter 5 and have the 6 dB bandwidth required by GB/T 6113.101—2016-Chapter 4.

A balanced dipole antenna should be used for antennas below 1 GHz. When the frequency is equal to or higher than 80 MHz, the length of the antenna should be the resonant length; when the frequency is lower than 80 MHz, the length should be equal to the resonant length of 80 MHz. For details, see the provisions of Chapter 4 of GB/T 6113.104—2008. (Note: Other antennas can be used as long as the measurement results are related to the results of the balanced dipole antenna and have acceptable accuracy.)

The antenna is used to measure the radiation field. The horizontal distance of the antenna from the EUT frame should meet the requirements of GB/T 6113.101—2008-Chapter 6. The EUT's border is delimited by an imaginary line that reflects the simple geometry of the EUT. All cables between the ITE system and the ITE to which they are connected should be located within this frame.

Note: If there is a strong ambient level or the measurement cannot be made at a distance of 10 m for other reasons, the measurement of the Class B EUT can be performed at a closer distance, for example 3 m. In order to determine whether the requirements are met, the inverse ratio factor of 20 dB/dec the distance is used to normalize the measurement data to the specified measurement distance. At around 30 MHz, attention should be paid to the effects of near-field effects when measuring from a large EUT 3 m. Distance from the antenna to the ground plane should be adjusted within the range of 1–4 m to obtain the maximum indication value.

During the measurement, the azimuth between the antenna and the EUT should be changed to find the maximum field strength. To this end, a method of rotating the EUT can be employed. If this is difficult, the position of the EUT can be fixed and the antenna can be measured around the EUT.

During the measurement process, in order to find the maximum field strength value, the antenna should be changed to horizontal or vertical polarization in sequence with respect to the EUT.

2. Measure Composition

The size of the anechoic chamber and the selection of the RF absorbing materials are mainly determined by the external dimensions and test requirements of the EUT, which are divided into 1 m, 3 m, or 10 m, which refers to the distance from EUT to the antenna center. Various standard metering agencies use anechoic chambers for radiated EMI noise testing. The following mainly introduces the radiation emission measurement in the 3 m method anechoic chamber and the 10 m method anechoic chamber.

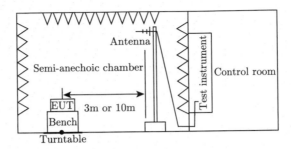

Fig. 4.12 Radiation emission test

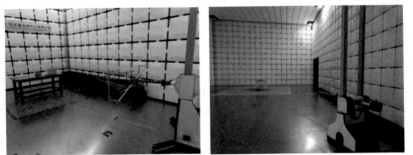

(a) 3m anechoic chamber test environment (b) 10m anechoic chamber test environment

Fig. 4.13 Test environment

Figure 4.12 shows the radiation emission test layout according to the requirements of CISPR16, CISPR11, CISPR13, CISPR15, CISPR22, and EN55022. In the radiation emission test, EUT is placed inside the semi-anechoic chamber on the turntable. A 360° rotation is performed to find the maximum radiation point with the receiving antennas being vertically polarized and horizontally polarized, respectively. After the radiation signal is received by the receiving antenna, it is transmitted through the cable to the receiver in the semiradio outdoor, and the waveform is obtained and compared with the limit.

Figure 4.13a is a 3 m semi-anechoic chamber test environment. It mainly consists of 3 m anechoic chamber, operation control room, and power amplifier room. It is for 3 m distance radiation immunity test and radiation emission compliance testing. The complete darkroom system can cover CISPR, EN, FCC, GB, GJB, and other standards, with the ability to upgrade to full-wave anechoic chamber. It can be widely used in electromagnetic compatibility testing of IT products, household appliances, medical equipment, auto parts, small military electronic products, industrial electronic products, and communication fields. Special products can be tested with special interfaces, such as water meters, gas meters, and energy meters.

Figure 4.13b shows the test environment of the 10 m anechoic chamber. The test environment consists mainly of the main anechoic chamber, control room, and power amplifier room for the compliance test of 10 m radiation emission. The 10 m anechoic

Table 4.2 Main performance comparison table of darkroom

Items	3 m anechoic chamber	10 m anechoic chamber
Radiation emission test	Meet the requirements of EN 50022, EN 50147-2, GJB 151A and GJB 152A	Meet the requirements of EN 50022, EN 50147-2, GJB 151A and GJB 152A
Quiet zone size	2 m in diameter and 2 m in height	4 m diameter cylinder with a height of 2 m (or a larger size quiet zone as needed)
Shielding effectiveness	Meet the standard EN 50147-1 or better	Meet the standard EN 50147-1 or better
Normalized site attenuation (NSA)	±3.5, ±3.0, ±2.5 dB, etc., available in CISPR16-1-4 and ANSI C63.4 standards from 30 MHz to 1 GHz	±3.5, ±3.0, ±2.5 dB, etc., available in CISPR16-1-4 and ANSI C63.4 standards from 30 MHz to 1 GHz
Field strength uniformity (FU)	Measured according to the requirements of IEC 61000-4-3, within 0–6 dB	Measured according to the requirements of IEC 61000-4-3, within 0–6 dB
Site voltage standing wave ratio (SVSWR)	Not more than 6 dB according to CISPR16-1-4 standard in the range of 1~18GHz	Not more than 6 dB according to CISPR16-1-4 standard in the range of 1–18 GHz
Ambient background noise (ABN)	10 dB lower than the limit specified by the CISPR22 standard	10 dB lower than the limit specified by the CISPR22 standard

chamber is the most popular standard test environment, mainly to meet the testing requirements of large-scale electronic systems and the industry test standards for some electronic products. It can also be tested by the 3 m anechoic chamber, and the quiet zone can be expanded. It is widely used in electromagnetic products such as IT products, household appliances, medical equipment, automobile vehicles and parts, large military electronic systems, industrial electronic systems, and communication fields, up to 5 m or even 8 m in diameter or larger. An electromagnetic compatibility test project for road vehicles can be completed by adding a drum system.

Table 4.2 shows the comparison of the main performances of the 3 m anechoic chamber and the 10 m anechoic chamber, which shows the comparison of advantages and disadvantages of them.

4.2.3 Radiation Emission Limitation

Commonly used radiated EMI standards include GB 9254—2016, GB 4824—2019, GB/T 18387—2017, EN 55022, FCC Part 15, CISPR 22, CISPR 25, etc.

When measuring the measurement distance R, the EUT should meet the limits of Table 4.3. If the reading shown on the measurement receiver fluctuates around the limit, the reading time should be no less than 15 s and the highest reading should be recorded, and the isolated instantaneous high value is ignored.

Table 4.3 Radiated EMI noise limits of the EUT at the measurement distance R (below 1 GHz, GB 9254)

Frequency range/MHz	$R = 3$ m		$R = 10$ m	
	Quasi-peak limit/(dB μV/m)		Quasi-peak limit/(dB μV/m)	
	A	B	A	B
30–230	50	40	40	30
230–1000	57	47	47	37

Note 1. Lower limits should be used at the transition frequency (230 MHz)
Note 2. When interference occurs, other regulations are allowed

The test site should be able to distinguish between radiated EMI noise and environmental noise from the EUT. It can be determined by measuring the ambient noise level (the EUT is not operational): The noise level should be at least 6 dB below the limit specified in Table 4.3. When the combined results of both ambient noise and radiated EMI noise of the EUT do not exceed the specified limits, it is not necessary to require the ambient noise level to be 6 dB below the specified limit. In this case, the emission of the source can be considered to meet the requirements of the specified limit; and when the combined results of the ambient noise and the radiated EMI noise of the EUT exceed the specified limit, it cannot be determined that the EUT does not meet the limit requirement. Unless each frequency point corresponding to the limit meets:

(1) The ambient noise level is at least 6 dB lower than the EUT radiated EMI noise plus the ambient noise level;
(2) The ambient noise level is at least 4.8 dB below the specified limit.

The radiated EMI noise test environment includes an open field, an anechoic chamber, and a giga hertz transverse electromagnetic cell (GTEM). Measurements are made with a measurement receiver with a quasi-peak detector in the frequency range of 30 MHz to 1000 MHz. To save test time, peak measurements can be used instead of quasi-peak measurements. In case of dispute, the measurement result of the quasi-peak measurement receiver shall prevail.

4.2.4 Requirements of the Device Under Test

The layout of the desktop device under test is shown in Fig. 4.14. The specific requirements are as follows:

(1) The interconnection I/O cable should be greater than 40 cm from the ground;
(2) In addition to the actual load connection, the device under test can also be connected to the analog load, but the simulated load should be able to conform to the impedance relationship and be able to represent the actual situation of the product application;

Fig. 4.14 Desktop device layout

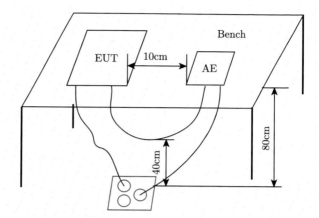

(3) The power cord of the device under test and the auxiliary device AE are directly plugged into the socket of the ground, and the socket should not be extended;
(4) The distance between the device under test and the auxiliary device AE is 10 cm;
(5) If the cable of the device under test is relatively large, it should be carefully smoothed, processed separately, and recorded in the test report, in order to obtain the weight of retesting.

Figure 4.15 shows the layout of the floor-standing device under test. The specific requirements are as follows:

(1) The I/O interconnection between the cabinets should be placed naturally. If it is too long, it can be tied into a bundle of 30–40 cm.
(2) The device under test is placed on the metal plane and is insulated from the metal plane by about 10 cm; the cable connected to the analog load or the anechoic chamber outdoor port should be insulated with the metal plane;
(3) The power cable of the device under test is too long and should be tied into a wire harness with a length of 30–40 cm, or shortened to just enough;
(4) If there are more cables of the device under test, it should be carefully smoothed, processed separately, and recorded in the test report to obtain repeatability of the test.

Some devices are not pure desktop devices or floor-standing devices. It may be a combination of the two. The experimental layout should be as shown in Fig. 4.16.

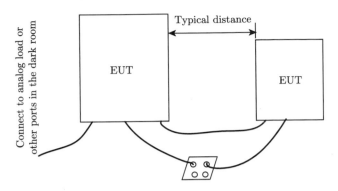

Fig. 4.15 Floor-standing equipment layout

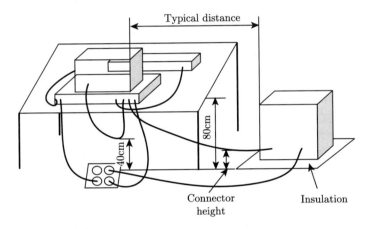

Fig. 4.16 Combined device layout

4.3 Diagnosis of Radiated EMI Noise

4.3.1 Principles of Radiated EMI Noise Diagnosis

Similar to the measurement of conducted EMI noise, the far-field radiation test result of the EUT obtained by the radiation EMI measurement in anechoic chamber (which is the total noise of the far-field radiation) does not separate the radiated common mode from differential mode noise. However, the generated mechanism of radiated common mode noise and radiated differential mode noise is also different, so there are also differences in the suppression methods. In order to effectively suppress radiated EMI noise, it is necessary to first diagnose the radiated EMI noise.

1. **Diagnosis of Differential Mode Radiation**

   1. Basic circuit
      Basic circuit of differential mode radiation is shown in Fig. 4.17.

Fig. 4.17 Basic circuit of
differential mode radiation

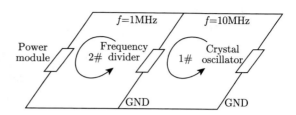

2. Basic characteristics

The far-field radiation electromagnetic field of the current loop is obtained by

$$E_\phi \approx \frac{I\mathrm{d}Sk^3}{4\pi\omega\varepsilon_0 r} \sin\theta e^{-jkr} \tag{4.44}$$

Therefore, the amplitude of the differential mode radiation is obtained by:

$$E_{\mathrm{DM}} = \left|E_\phi\right| = \frac{I\mathrm{d}Sk^3}{4\pi\omega\varepsilon_0 r} \sin\theta = 1.316 \times 10^{-14} \frac{f^2 I\mathrm{d}S}{r} \sin\theta \tag{4.45}$$

In fact, the radiated electromagnetic field generated from the digital device is measured by the receiving antenna and has a ground plane. The highest electromagnetic field measured by the antenna is accepted when the following two conditions are met:

(1) The accepted antenna in the direction of the electromagnetic field is $\theta = 90°$.
(2) The electromagnetic field at the ground plane is in the same phase as the electromagnetic field toward the receiving antenna. Therefore, the measured amplitude is twice of which is in free space.

In the above two cases, the amplitude of the differential mode radiation can be simplified as follows:

$$E_{\mathrm{DM}} = 2.632 \times 10^{-14} \frac{f^2 I\mathrm{d}S}{r} \tag{4.46}$$

In the above formula, $\mathrm{d}S$ is the area of the loop, the unit is square meter, f is the frequency, the unit is Hertz, I is the current, the unit is ampere, r is the distance from the ring to the receiving antenna, and the unit is meter.

It can be obtained that the radiation field strength is proportional to the loop area. The higher the loop frequency, the greater the influence on the differential mode radiation is.

3. Diagnosis

(1) Decrease the area of the main frequency loop 1 in Fig. 4.17, and the measured radiation field strength is greatly reduced. It reveals that the radiation field strength has a significant relationship with the main frequency loop area.

Fig. 4.18 Basic circuit of common mode radiation

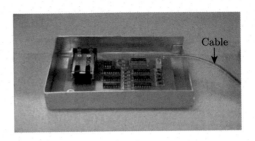

(2) The area of loop 1 is constant, and the area of loop 2 is reduced. The measured radiation field strength is reduced, but the change is not obvious. It can be concluded that although the loop area is reduced, the amplitude formula of differential mode radiation is clear. It is shown that the radiation field strength is proportional to the square of the frequency, and loop 2 is a low-frequency loop, so the influence on the differential mode radiation is small.

2. Diagnosis of Common Mode Radiation

1. Basic circuit
   Basic circuit of common mode radiation is shown in Fig. 4.18.
2. Basic characteristics

$$E_{CM} = |E_\theta| = \frac{I dl k^2}{4\pi\omega\varepsilon_0 r} \sin\theta = 6.28 \times 10^{-7} \frac{I dl f}{r} \sin\theta \qquad (4.47)$$

The above formula is a radiated electromagnetic field in free space. But when the electric field at the ground plane is in phase with the electric field toward the electric field, the electric field is the largest, which is $\theta = 90°$.
Simplify the above formula to get the highest radiation electric field:

$$E_{CM} = 1.256 \times 10^{-6} \frac{I dl f}{r} \sin\theta \qquad (4.48)$$

dl is the length of the cable in unit of meter. f is frequency in unit of Hertz. I is current in unit of ampere. r is a distance from the line to the receiving antenna in unit of meter.
Common mode radiation can be controlled by reducing the current, reducing the frequency, and shortening the length of the cable.
3. Diagnosis
   The length of the external cable connected to the main frequency of the circuit is reduced, and the measured common field radiation field strength is greatly reduced. It reveals that the common mode radiation is determined by the cable length at the highest-frequency common mode current.

4.3.2 Near-Field Wave Impedance Measurement for Radiated EMI Noise Diagnosis

In this book, EMI noise models of electric dipole common mode radiation and magnetic dipole differential mode radiation are established, respectively. Because the common mode and differential mode noise have the same characteristics in the far-field region, that is, there is a one-to-one correspondence between the radiated electric field and the radiated magnetic field. Therefore, this section focuses on the electromagnetic propagation characteristics of common mode and differential mode noise in the near-field region, and a diagnostic method of radiated EMI noise mechanism based on near-field wave impedance measurement is designed.

In order to reveal the electromagnetic transmission characteristics in the near and far fields, wave impedance is defined as

$$Z = \frac{E}{H} \tag{4.49}$$

In the formula, Z is wave impedance, E is electromagnetic strength, and H is magnetic field strength.

In the far field, Z is $120\pi\Omega$, while in the near field, Z is not a constant, but is related to the test distance.

1. Electric Dipole Common Mode Radiated EMI Noise

For near-field common mode radiation, it can be obtained from Maxwell's formulae and Eq. (4.49).

$$\begin{cases} H_\phi = \dfrac{I \mathrm{d}l k^2}{4\pi}\left[-\dfrac{1}{jkr} + \dfrac{1}{(kr)^2}\right]\sin\theta \mathrm{e}^{-jkr} \\[2mm] E_\theta = \dfrac{I \mathrm{d}l k^3}{4\pi\omega\varepsilon_0}\left[\dfrac{-1}{j\,(kr)} + \dfrac{1}{(kr)^2} + \dfrac{1}{j\,(kr)^3}\right]\sin\theta \mathrm{e}^{-jkr} \\[2mm] E_r = \dfrac{I \mathrm{d}l k^3}{2\pi\omega\varepsilon_0}\left[\dfrac{1}{(kr)^2} + \dfrac{1}{j\,(kr)^3}\right]\cos\theta \mathrm{e}^{-jkr} \end{cases} \tag{4.50}$$

In the formula, H is the magnetic field strength, E is the electric field strength, $I\mathrm{d}l$ is the dipole moment, k is the wave vector, the mode represents the wave number, the direction represents the direction of wave propagation, r is the test distance, ε_0 is the vacuum dielectric constant, and ω is the angular frequency.

In the common mode radiated field of an electric dipole, the relationship between the magnetic field strength and the electric field strength and the test distance is as follows.

$$H \propto \frac{1}{r^2} \quad E \propto \frac{1}{r^3} \tag{4.51}$$

It can be seen from Eq. (4.51) that the electric field intensity of common mode EMI noise is inversely proportional to the cube of the test distance, while the magnetic field intensity is proportional to the square of the test distance. Therefore, the near-field wave impedance of common mode radiation is

$$Z_{CM} = \frac{E}{H} \propto \frac{1}{r} \tag{4.52}$$

It is not difficult to find from Eq. (4.52) that the near-field wave impedance of common mode radiation is inversely proportional to the test distance and exhibits a high impedance (greater than $120\pi\Omega$). On the other hand, the common mode radiated field is related to the length of the equivalent short straight antenna. The longer the equivalent length, the stronger the common mode radiated field is.

2. Magnetic Dipole Differential Mode Radiated EMI Noise

For near-field differential mode radiation, it can be obtained from Maxwell's formulae and Eq. (4.49).

$$\begin{cases} H_\theta = \dfrac{I\,dSk^3}{4\pi} \left[-\dfrac{1}{kr} - \dfrac{1}{j\,(kr)^2} + \dfrac{1}{(kr)^3} \right] \sin\theta e^{-jkr} \\[2mm] H_r = \dfrac{I\,dlk^3}{2\pi} \left[\dfrac{-1}{j\,(kr)^2} + \dfrac{1}{(kr)^3} \right] \cos\theta e^{-jkr} \\[2mm] E_\phi = \dfrac{I\,dSk^4}{4\pi\varepsilon_0\omega} \left[\dfrac{1}{kr} + \dfrac{1}{j\,(kr)^2} \right] \sin\theta e^{-jkr} \end{cases} \tag{4.53}$$

In the formula, $I\,dS$ is the magnetic dipole moment.

According to Eq. (4.53), the relationship between magnetic field intensity and electric field intensity and test distance in magnetic dipole differential mode radiated field is as follows:

$$H \propto \frac{1}{r^3} \quad E \propto \frac{1}{r^2} \tag{4.54}$$

It can be seen from Eq. (4.54) that the differential mode electric field intensity of radiated EMI noise is proportional to the square of the test distance, while the magnetic field intensity is inversely proportional to the cube of the test distance.

3. Near-field wave impedance noise diagnosis

Therefore, the near-field impedance of differential mode radiation is

$$Z_{DM} = \frac{E}{H} \propto r \tag{4.55}$$

It is not difficult to find from Eq. (4.55) that the near-field wave impedance of differential mode radiation is proportional to the test distance and exhibits a low impedance (less than $120\pi\Omega$). On the other hand, the differential mode radiated field

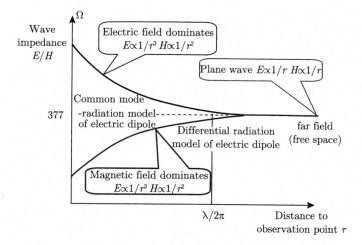

Fig. 4.19 Relation between wave impedance and test distance

is related to the area of the equivalent annular antenna. The larger the equivalent area, the greater the intensity of the differential mode radiation field is.

The transmission characteristics of common mode and differential mode radiated EMI noise can be known from Eqs. (4.52) and (4.55), as shown in Fig. 4.19.

Through near-field test, radiated EMI noise can be extracted from electric dipole common mode radiation and magnetic dipole differential mode radiation from Fig. 4.19, Eqs. (4.52) and (4.55), thus the mechanism diagnosis of radiated EMI noise can be realized.

References

1. Mardiguian M. Controlling radiated emissions by design. Hoboken: Wiley; 1992. p. 684.
2. Mills JP. Electromagnetic interference reduction in electronic systems. New Jersey: Prentice Hall; 1993.
3. Montrose MI. EMC and the printed circuit board: design, theory, and layout made simple. Piscataway, NJ: IEEE Press; 1998.
4. Johnson HW, Graham M. High-speed digital design: a handbook of black magic. New Jersey: Prentice-Hall; 1993.
5. See KY, Freeman FM. Rigorous approach to modelling electromagnetic radiation from finite-size printed circuit structures. IEE Proc Microwaves Antennas Propag. 1999;146(1):29–34.
6. See KY, Ma JG. Electromagnetic radiation from printed-circuit structure with extended conductor. Microwave Opt Technol Lett. 2000;27(2):118–20.
7. See KY. Evaluation of PCB layout guidelines for controlling radiated EMI. J Inst Eng Singap: Electron Comput Eng. 2001;41(1):6–11.
8. Chua EK, See KY. Full-wave equivalent circuit model for 2D conductor. Electron Lett. 2003;39(19):1367–9.

9. See KY, Chua EK. SPICE-compatible full-wave equivalent circuit model for interconnect structures. In: Proceedings of the 5th electronics packaging technology conference, Singapore, 2003, vol. 12. p. 168–70.

10. See KY, Chua EK, Zhao Y. Modeling common-mode radiation from high-speed PCB using method of moment. In: Proceedings of the 5th electronics packaging technology conference, Singapore, 2003, vol. 12. p. 523–5.

11. Kee LK, See KY, Chin R. Modeling of power/ground noise of high-speed digital circuit using SPIC. Integr Circuits Devices Syst. 2004;9:357–60.

Chapter 5
Radiated EMI Noise Suppression Methods and Cases Study

5.1 Suppression Principle of Radiated EMI Noise

5.1.1 Common Mode Suppression Principles

Common mode radiation is caused by the voltage drop in the grounding circuit, and some parts have a high-potential common mode voltage, when the external cables are connected to these parts, common mode current is generated under the excitation of common mode voltage, which becomes the antenna of radiated electric field. This is mostly due to the existence of voltage drop in the grounding system. Common mode radiation usually determines the radiation performance of products.

It can be seen from Chap. 4 that the common mode radiation is mainly radiated from the cables and can be simulated by a short monopole antenna with a length of less than 1/4 wavelength excited by the ground voltage. The current on the ideal antenna is uniform, and the actual antenna top current tends to be 0. The actual cable has one device connected to the other end, which is equivalent to a capacitive load antenna, that is, the end of the antenna relates to a metal plate. At this time, a uniform current flows through the antenna, and the antenna is pointed to the maximum field strength, and the maximum field strength is obtained by calculation formula:

$$E_{CM} = 12.6 \times 10^{-7} \frac{f l I_{CM}}{r} \tag{5.1}$$

where f is the signal frequency, I_{CM} is the common mode current in the circuit, l is the length of the radiated circuit wire, and r is the test distance. It can be seen from the formula that the common mode radiation is proportional to the length l of the cable, the frequency f of the common mode current, and the common mode current intensity I, and is inversely proportional to the test distance r. The equivalent circuit of the common mode radiation model is shown in Fig. 5.1. Among them, U_{CM} is the common mode radiation voltage, I_{CM} is the common mode radiation current, and Z_{CM} is the line equivalent impedance.

© Science Press 2021
Y. Zhao et al., *Electromagnetic Compatibility*,
https://doi.org/10.1007/978-981-16-6452-6_5

Fig. 5.1 Equivalent circuit
of common mode radiation
model

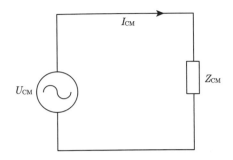

5.1.2   Differential Mode Suppression Principles

Assuming that the differential mode radiation model is a small loop current, considering the total reflection of the ground plate and the direction of the maximum radiated field strength, the expression of the magnetic dipole field strength at the far field can be deduced as follows.

$$E_{DM} = 2.63 \times 10^{-14} \left(f^2 A I_{DM} \right) \left(\frac{1}{r} \right) \qquad (5.2)$$

where f is the signal frequency, I_{DM} is the differential mode current in the circuit, A is the circuit loop area, and r is the test distance. The differential mode radiated field is proportional to the square of the signal frequency, the signal current loop area, and the differential mode current, and is inversely proportional to the test distance.

   Ways to reduce differential mode radiation:

According to the calculation formula of differential mode radiation, the method of reducing differential mode radiation can be obtained as:

(1) Reducing the operating frequency of the circuit;
(2) Reducing the area of the signal loop;
(3) Reducing the strength of the signal current.

In PCB circuit, high-speed processing speed is guaranteed by the high clock frequency, so limiting the operating frequency of the system is sometimes not allowed. The limiting frequency refers to reducing unnecessary high-frequency components. Mainly refers to the frequency above $1/\pi t_r$ frequency. The strength of the signal current cannot be reduced casually, but sometimes the buffer can reduce the drive current on the long line. The most realistic and effective method is to control the area of the signal loop. The radiation of the loop can be effectively reduced by reducing the signal loop area.

5.2   Suppression Methods and Analysis of Radiated EMI Noise

5.2.1   Common Mode Suppression Method

The common mode radiation is proportional to the frequency f of the common mode current, the common mode current I, and the length l of the antenna (cable). Therefore, reducing the common mode and differential mode radiation should reduce the frequency f, the current I, and the length l, and limiting the common mode current I is the basic method to reduce the common mode radiation. To do this, the following process should be achieved:

(1) Minimizing the source voltage that excites this antenna, that is, the ground potential;
(2) Providing a high common mode impedance in series with the cable, that is, increasing the common mode choke;
(3) Bypassing the common mode current to ground;
(4) Reducing the length of the cable;
(5) Cable shielding layer terminates with shielded shell at 360°.

The ground plane can be used to effectively reduce the ground potential in the grounding system. In order to bypass the common mode current to ground, the ground plane of the printed circuit board can be separated from the connector as a "noise-free" input/output ground. To avoid contamination of the input/output ground, only allow the decoupling capacitor of the input and output lines and the shield of the external cable are connected to the "noise-free" ground, and the inductance of the decoupling loop is as small as possible. In this way, the common mode current of the printed circuit board carried by the input/output line is bypassed to the ground by the decoupling capacitor, and the external interference is also bypassed to the ground by the decoupling capacitor when it has not reached the component area, thereby protecting the internal components work properly. The two wires are wound in the same direction on the ferrite magnetic ring to form a common mode choke coil, the DC and low-frequency time difference mode current can pass, but the high-frequency common mode current shows a large impedance, and the increase of Z_{CM} results in the decrease of common mode radiation current I_{CM}, which suppresses the common mode radiation field intensity.

Most of the existing noise suppression schemes reduce the radiated noise by reducing the high-frequency noise current in the PCB circuit, but the radiated noise source cannot be weakened in essence, the signal-to-noise ratio is low, and the noise suppression effect is also poor. According to the radiation mechanism diagnosis results, this section weakens the common mode radiated field by using the magnetic core, reducing the length of the transmission cable, and improving the grounding to reduce the reflection potential of the ground point, which provide theoretical basis for future engineering applications of radiated EMI noise suppression.

1. Devices

(1) Magnetic Beads

Magnetic beads are designed to suppress high-frequency noise and spikes on signal lines, power lines, and the ability to absorb electrostatic pulses. Magnetic beads are used to absorb ultrahigh frequency signals, such as some RF circuits, PLL, oscillating circuits, and ultrahigh frequency memory circuits (DDRSDRAM, RAMBUS, etc.), which need to add magnetic beads to the input part of the power supply, but the inductance is an energy storage device and used in LC oscillator circuits, low-frequency filter circuits, etc., whose application frequency rarely exceeds 50 MHz. Magnetic beads have high resistivity and permeability, equivalent to series connection of resistance and inductance, but both resistance and inductance vary with frequency. The magnetic beads are shown in Fig. 5.2.

The main raw material of the magnetic beads is ferrite. Ferrite is a ferromagnetic material of a cubic lattice structure. The ferrite material is iron–magnesium alloy or iron–nickel alloy, and its manufacturing process and mechanical properties are like those of ceramics, and the color is gray-black. One type of magnetic core that is often used in electromagnetic interference filters is ferrite material. This material is characterized by a very high frequency loss and a high magnetic permeability, which minimizes the capacitance generated by the high frequency and high impedance between the coil windings of the inductance. The most important performance parameters for ferrite for suppressing electromagnetic interference are magnetic permeability μ and saturation magnetic flux density B_s. The magnetic permeability μ can be expressed as a complex number, the real part constitutes inductance, and the imaginary part represents loss, which increases as the frequency increases. Therefore, its equivalent circuit is a series circuit consisting of an inductance L and a resistance R, both L and R being a function of frequency. When a wire passes through such a ferrite core, the inductance impedance increases formally with the increase of frequency, but the mechanism is completely different at different frequencies.

In the low-frequency band, the impedance is composed of the inductive reactance of the inductance. R is very small at low frequency, the magnetic permeability of the magnetic core is high, so the inductance is large, L plays a major role, and electromagnetic interference is suppressed by reflection; and at this time, the core loss is small, the whole device is a low loss, high Q characteristic inductance, which

Fig. 5.2 Magnetic beads

easily causes resonance, so in the low-frequency band, sometimes there may be interference enhancement after using ferrite beads.

In the high-frequency band, the impedance is composed of a resistance component. As the frequency increases, the magnetic permeability of the magnetic core decreases, resulting in a decrease of the inductance and a decrease in the inductive component. However, the loss of the core increases and the resistance component increases, which causes the total impedance to increase, and when the high-frequency signal passes through the ferrite, the electromagnetic interference is absorbed and converted into heat energy to dissipate.

Ferrite suppression components are widely used in printed circuit boards, power lines, and data lines. High-frequency interference can be filtered out by adding a ferrite suppression component to the input end of the power strip of the printed board. Ferrite core inductor or magnetic beads are designed to suppress high-frequency interference and spike interference on signal lines and power lines; they also can absorb electrostatic discharge pulse interference. The numerical value of the two components is proportional to the length of the magnetic beads, and the length of the magnetic beads has a significant effect on the suppression effect. The longer the length of the magnetic beads, the better the suppression effect.

The high-frequency equivalent model of a series ferrite bead is shown in Fig. 5.3a, equivalent to an inductor and a resistor in series. Ferrite magnetic beads are available in both single-turn and multi-turn, and the increase in the impedance of the multi-rhenium ferrite bead is proportional to the square of the number of turns. However, the increase in the number of turns also increases the inter-cell capacitance, which reduces the high-frequency impedance of the ferrite. Therefore, it is suitable for low-frequency situations. As shown in Fig. 5.3b, at the oscillation frequency, the ferrite bead is inductive and is used to form a low-pass L-type filter that attenuates the EMI noise generated by the high-frequency oscillator without reducing it. Figure 5.3c is a magnetic bead equivalent model diagram. Figure 5.3d shows the practical application of ferrite beads.

The circuit symbol of the magnetic bead is the inductance. The magnetic bead is used in the circuit function, the magnetic bead and the inductance share the same principle, but the frequency characteristics are different. The magnetic beads are composed of an oxygen magnet. The magnetic beads have a greater hindrance to the high-frequency signal. As shown in Fig. 5.4, the Type-43 magnetic beads have the highest impedance at 300 MHz, and the magnetic beads have the best suppression effect. It has a much lower impedance than inductance at low frequency.

(2) Ferrite Core Inductor

The ferrite core inductor is a ring-shaped magnet. The ferrite core inductor is a commonly used anti-interference component in electronic circuits and has a good inhibitory effect on high frequency noise. It is generally made of ferrite material (Mn–Zn). The ferrite core inductor has different impedance characteristics at different frequencies. Generally, the impedance is small at low frequency, and the impedance of the ferrite core inductor rises sharply as the signal frequency increases. Figure 5.5 shows the ferrite core inductor.

(a) High frequency equivalent model of series ferrite beads

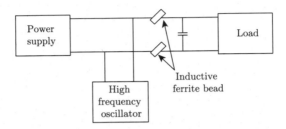

(b) L-type filter with ferrite beads to prevent high frequency oscillator noise from entering the load

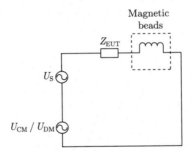

(c) Magnetic bead equivalent model

(d) Application of ferrite beads

Fig. 5.3 Ferrite beads

Fig. 5.4 Type-43 magnetic
bead characteristics

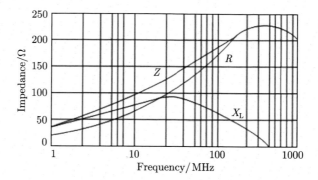

Fig. 5.5 Ferrite core
inductor

The magnetic permeability μ is the most important parameter of the ferrite core inductor. Due to the application, the actual magnetic permeability may not be the initial permeability μ_i, but the effective permeability μ_e, the maximum permeability μ_m or the incremental permeability μ_Δ. In general, the other three parameters of the material with high initial permeability are also high. Therefore, the initial permeability is usually used as the basic parameter of the magnetic material, which is [1]:

$$\mu_i = \frac{1}{\mu_0} \lim_{H \to 0} \frac{\Delta B}{\Delta H} \tag{5.3}$$

The ferrite core inductor operating in the high-frequency weak field always has a high initial permeability μ_i of the material. Because of the higher the initial permeability, the smaller the component volume can be at the same inductance requirement. At the same time, the influence of stray magnetic fields such as magnetic flux leakage can also be reduced.

According to the definition of physics, the alternating magnetic field H and the magnetic induction B are [1, 2]

$$\begin{cases} H = H_m e^{j\omega t} \\ B = B_m e^{j(\omega t - \sigma)} \end{cases} \tag{5.4}$$

According to permeability [2]:

$$\mu = \frac{B}{\mu_0 H} \qquad (5.5)$$

Combining Eqs. (5.4) and (5.5):

$$\mu = \frac{B_m e^{j(\omega t - \delta)}}{\mu_0 H_m e^{j\omega t}} = \frac{B_m}{\mu_0 H_m} e^{-j\delta} = \frac{B_m}{\mu_0 H} (\cos \delta - j \sin \delta) \qquad (5.6)$$

Recorded as

$$\begin{cases} \mu' = \dfrac{B_m}{\mu_0 H} \cos \delta \\ \mu'' = \dfrac{B_m}{\mu_0 H} \sin \delta \end{cases} \qquad (5.7)$$

Then the magnetic permeability of the ferrite core inductor can be expressed as [2]:

$$\mu = \mu' - j\mu'' \qquad (5.8)$$

In the formula, the real part μ' constitutes the inductance of the ferrite core inductor; the imaginary part μ'' constitutes the ferrite core inductor impedance.

When winding a toroidal ferrite core inductor, the coil current is I, the magnetic field strength H generated by the coil in the ferrite core inductor is

$$H = \frac{NI}{2\pi r} \qquad (5.9)$$

where r is ferrite core inductor radius and N is coil turns. The strength of the magnetic field is inversely proportional to the radius of the ferrite core inductor, indicating that the geometry of the core has an effect on the strength of the magnetic field [3]. An ideal ferrite core inductor has been uniformly magnetized, having an effective magnetic path length I_e and an effective sectional area A_e. For the toroidal core shown in Fig. 5.6.

$$\begin{cases} I_e = \dfrac{2\pi \ln \dfrac{r_2}{r_1}}{\dfrac{1}{r_1} - \dfrac{1}{r_2}} \\ A_e = \dfrac{h \left(\ln \dfrac{r_2}{r_1} \right)^2}{\dfrac{1}{r_1} - \dfrac{1}{r_2}} \end{cases} \qquad (5.10)$$

where h is the core height; $r_1 (r_2)$ is the inner (outer) radius of the toroidal ferrite core, and it is known that the effective cross-sectional area is proportional to the core height.

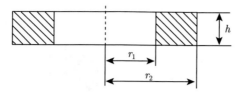

Fig. 5.6 Ring section

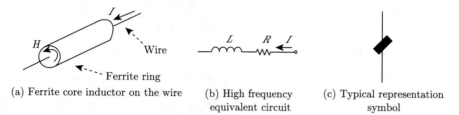

(a) Ferrite core inductor on the wire   (b) High frequency   (c) Typical representation
equivalent circuit   symbol

Fig. 5.7 Ferrite core inductor on the wire

In engineering practice [3, 4], I_e/A_e is usually called the core size constant, which is represented by C_1.

$$C_1 = \frac{I_e}{A_e} = \frac{2\pi}{h \ln\left(\dfrac{r_2}{r_1}\right)} \tag{5.11}$$

In the formula, C_1 reflects the influence of the shape of the core on the inductance of the core coil, and the unit is m^{-1}. At the same core volume, a core with a small core size constant C_1 can achieve a greater inductance. It can be seen from Eqs. (5.9) and (5.10) that the thickness of the core is proportional to the length of the core, and by doubling the core height h, the core volume is doubled and the inductance is doubled. It can be seen from the calculation that the change of the inner (outer) diameter of the magnetic core increases the inductance of the magnetic core by less than 50%. Therefore, a longer ferrite core inductor is preferred in the filter selection.

Figure 5.7a shows a small cylindrical ferrite bead mounted on a wire; Fig. 5.7b shows equivalent circuit at high frequency: an inductor in series with a resistor, the values of the impedance and inductance depend on the frequency, the resistance is derived from the high-frequency hysteresis loss of the ferrite material; Fig. 5.7c shows the common representation of the ferrite bead.

According to the equivalent circuit shown in Fig. 5.7b, the equivalent impedance Z of the ferrite core is a function changed with frequency [4].

$$Z = X_L + R_S = j\omega L_S + R_S \tag{5.12}$$

Ferrite provides an inexpensive way to couple high-frequency impedance into a circuit without loss of energy in direct current (DC) and without affecting low-

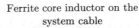

Ferrite core

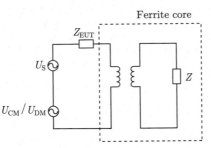

(a) Ferrite core inductor equivalent model   (b) Ferrite core inductor application example

Fig. 5.8 Suppression noise circuit based on ferrite core inductor

frequency signals. In general, ferrite can be considered as a high-frequency AC resistor, while the DC or low-frequency resistor has little or no resistance.

The equivalent circuit of the ferrite core inductor applied in the circuit is shown in Fig. 5.8. Z is the impedance of the ferrite core inductor, and Z_S and Z_L are the impedance of the source end and the impedance of the load end, respectively, and the ferrite core inductor is equivalent to the inductor.

According to Eq. (5.11), the relationship between the ferrite core inductor insertion loss and the impedance is [5]

$$I_L = 20 \lg \left| \frac{Z_S + Z_L + Z}{Z_S + Z_L} \right| \tag{5.13}$$

When Z_S and Z_L are constant, the impedance of the suppression element is proportional to the suppression effect. Magnetic loop impedance includes inductive reactance and resistance, both of which influence insertion loss. According to the derivation, at low frequency, the value of the ferrite core inductor is small, mainly because the inductive reactance acts, and the EMI signal is reflected and suppressed; at high frequency, the value of the ferrite decreases, the value of the ferrite core inductor increases, and the loss resistance plays a major role. The EMI signal is absorbed and converted into thermal energy.

(3) Common Mode Choke

The cable passes through a ferrite bead to form a common mode choke, and the cable can be wound around the ferrite core inductor as needed. The more the number of turns, the better the interference suppression effect on the lower frequency, and the weaker the noise suppression on the higher frequency. In actual engineering, the number of turns of the ferrite core inductor should be adjusted according to the frequency characteristics of the interference current. Generally, when the frequency band of the interference signal is wide, two ferrite core inductor can be placed on the cable, and each magnetic circle surrounds different turns, so that high-frequency interference and low-frequency interference can be suppressed at the same time.

From the mechanism of the common mode choke action, the greater the impedance, the better the interference suppression effect. The impedance of the common mode choke is from the common mode inductance $L_{CM} = j\omega L_{CM}$. It is easy to see from the formula that for a certain frequency of noise, the inductance of the ferrite core inductor is as large as possible.

The impedance formula of the winding [5, 6]:

$$Z = \frac{U}{I} = \frac{PLi_1 + PMi_2}{i_1} \tag{5.14}$$

The common mode current in the line is

$$I_c = \frac{1}{2}(I_1 + I_2) \tag{5.15}$$

Differential mode current is

$$I_d = \frac{1}{2}(I_1 - I_2) \tag{5.16}$$

In the circuit, when $I_1 = I_2 = I_c$, there is:

$$Z_{CM} = P(L + M) \tag{5.17}$$

I_c is common mode current.

When $I_1 = I_d$, $I_2 = -I_d$ The effect of differential mode current on series impedance is [6]

$$Z_{CM} = P(L - M) \tag{5.18}$$

I_d is differential mode current.

If the windings are symmetrical and all flux is in the core, which means that the flux of one winding is completely connected to the other windings, then $L = M$ and $Z_{DM} = 0$, so ideally, when $L = M$, the common mode choke has no effect on the differential mode current, but for the common mode current, it is necessary to connect a series of inductive reactance (impedance) $2L$ in the conductor.

The ferrite core inductor is a ferrite material, and the ferrite core inductor is stuck on the power line or the signal line. In the high-frequency case, as shown in Fig. 5.7, the ferrite core inductor has a mutual inductance effect on the cable, which is equivalent to connecting an inductor to the cable. If the live and neutral wires are stuck at the same time, it is equivalent to connecting a common mode inductance to the cable, increasing the common mode source impedance of the cable to suppress the noise current I_{CM} and reducing the radiation field strength E_{CM}, thereby suppressing radiation. In addition, according to the principle of electromagnetic induction, when the cable (power line) radiates high-frequency electromagnetic waves uninterruptedly, the strength of the magnetic field in the ferrite core inductor changes, thereby generating an induced current. At this time, if the ferrite core inductor is made of a

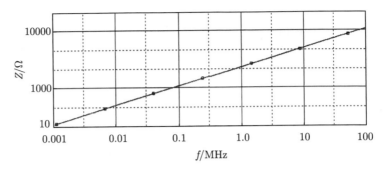

Fig. 5.9 Impedance of an ideal inductor

magnetic material with a high hysteresis coefficient and a low resistivity, the generated high frequency energy is converted into heat energy to be consumed, and the existing noise can be effectively suppressed.

(4) Inductance

Inductors are components that can be converted into magnetic energy and stored in a structure similar to a single-winding transformer. In an EMI filter, the inductor is usually wound into a coil. The core is a ferrite soft core. The core material with higher magnetic permeability than air can tightly bind the magnetic field around the inductor element. It is called a choke, a reactor, or a dynamic reactor.

Similar to a capacitor, the inductor in the low-frequency state has only a pure inductance value, but in the high-frequency state, the inductance characteristics deviate far from the rational-state characteristic. As shown in Fig. 5.9, the impedance of the inductor in an ideal case increases linearly with frequency at a rate of +20 dB/dec.

When the filter is applied to higher frequency, "parasitic parameters" becomes a very important factor. The high-frequency inductor is equivalent to the parallel connection of the ideal inductor L, equivalent parallel capacitance (EPC), and equivalent parallel resistance (EPR), where the EPC is the inductor coil due to "coil–coil." The parasitic capacitance generated between "coil–ground" and "coil–iron (magnetic) core" EPR is the equivalent resistance considering the eddy current loss in the inductor wire, core, and lead; considering the operating range of the inductor, the electromagnetic wavelength is much larger than the size, so the model can be represented by a centralized parameter.

As shown in Fig. 5.10a, in addition to the parasitic capacitance C_{lead} and the lead parasitic inductance L_{lead}, there is an equivalent winding resistance R_w and an equivalent core resistance R_p. Because the inductor is a passive component, the inductance value tends to far exceed the inductance of the lead inductance L_{lead}, so L_{lead} is often ignored. Similarly, the capacitance of the parasitic capacitance C_p is much larger than the capacitance of the lead parasitic capacitance C_{lead}, and the lead capacitance is often ignored. The simplified high-frequency model of the inductor is shown in Fig. 5.10b, c which is the equivalent model of the inductor.

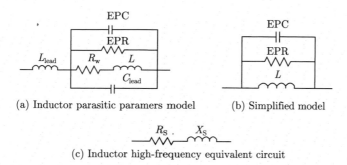

(a) Inductor parasitic paramers model   (b) Simplified model

(c) Inductor high-frequency equivalent circuit

Fig. 5.10 Equivalent high-frequency circuit model of inductive component

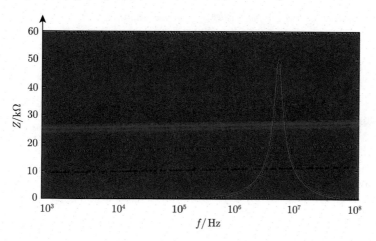

Fig. 5.11 Impedance of the inductor

R_S and X_S in Fig. 5.10c are [5, 6]

$$R_S = \frac{R_p L^2 \omega^2}{R_p^2 \left(1 - \omega^2 L C_p\right)^2 + \omega^2 L^2} \tag{5.19}$$

$$X_S = \frac{\omega R_p^2 L \left(1 - C_p L \omega^2\right)}{R_p^2 \left(1 - \omega^2 L C_p\right)^2 + \omega^2 L^2} \tag{5.20}$$

In the formula, $\omega = 2\pi f$.

Impedance in Fig. 5.10c is $Z_{S3} = R_S + jX_S$.

Figure 5.11 is a graph of the actual impedance characteristics and frequency of the inductor. The inductor cannot maintain single inductance characteristic at high frequency. This is because as the frequency increases, the absolute value of the inductance parasitic capacitance gradually increases to the absolute value of the inductance. Eventually, the inductive component behaves as capacitive.

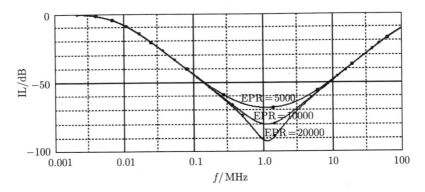

Fig. 5.12 Effect of equivalent parallel resistance on inductor insertion loss

5.2.2 Extraction of Inductance Equivalent Resistance

The relationship between the EPR of the inductor and the frequency is very complex, which increases the difficulty of EPR extraction. In this chapter, based on the empirical formula of frequency and core loss, it is derived that the equivalent parallel resistance is approximately linear with the frequency, which makes it easier to establish an accurate high-frequency model of the inductor.

As shown in Fig. 5.12, the equivalent parallel resistance EPR is taken as $5\,\mathrm{k\Omega}$, $10\,\mathrm{k\Omega}$, and $20\,\mathrm{k\Omega}$, respectively. By comparing the insertion loss of the inductor at this time, it is found that in the frequency band of about $1\,\mathrm{MHz}$, the larger the EPR, the better the filtering effect of the inductor.

The empirical formula for core loss is [4–6]

$$P_c = \eta f B_m^2 U_e \tag{5.21}$$

In the formula, P_c is the loss of the magnetic core, U_e is the volume of the magnetic core, η is the loss coefficient, f is the operating frequency, and B_m is the amplitude sensing intensity of the magnetic core.

According to the law of electromagnetic induction:

$$\mu_L = N A_e \frac{dB}{dt} \tag{5.22}$$

where A_e is the effective cross-sectional area of the magnetic circuit of the coil, N is the number of turns of the coil, and u_L is the excitation voltage. Then:

$$U_{rms} = \frac{N A_e \omega}{\sqrt{2}} B_m \tag{5.23}$$

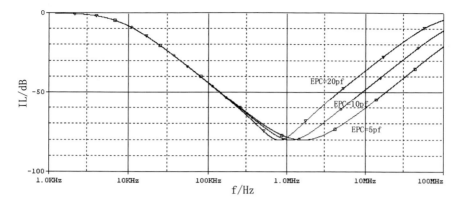

Fig. 5.13 Effect of equivalent capacitance on inductor insertion loss

Because $\omega = 2\pi f$, there is:

$$P_c = \frac{\eta l_e}{2\pi^2 N^2 A_e f} U_{rms}^2 \tag{5.24}$$

Finally, the equivalent resistance of the inductor is [6]

$$R_p = \frac{2\pi^2 N^2 A_e f}{\eta l_e} \tag{5.25}$$

5.2.3 Extraction of Inductance Equivalent Capacitance

As shown in Fig. 5.13, the equivalent parallel capacitance EPC is 5 pF, 10 pF, and 20 pF, respectively. By comparing the insertion loss of the inductor at this time, it can be found that the smaller the EPC of the inductor in the high-frequency band, the better the filtering effect of the inductor. Therefore, the extraction and suppression of EPC are very important.

It can be found from the circuit common sense that if L_c and C_p are connected in parallel in the circuit, parallel resonance phenomenon occurs at a certain frequency. If the resonance frequency is f_r, then C_p can be expressed as [6]

$$C_p = \frac{1}{(2\pi)^2 f_r^2 L_c} \tag{5.26}$$

where L_C is the inductance of the inductor at the resonant frequency f_r. Normally, the inductance value of the core follows the frequency change, so it is necessary to accurately obtain the C_p, and the inductance value must be measured first.

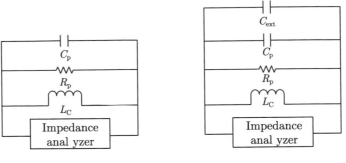

(a) Resonance measurement   (b) External and capacitance measurement

Fig. 5.14 Electrical connection diagram of impedance parasitic capacitance

However, due to the dispersion phenomenon, the magnetic permeability of the magnetic core changes with frequency, and it is generally difficult to directly measure the inductance value at the resonance frequency. At this point, the external capacitor is connected in parallel to solve the problem, as shown in Fig. 5.14. C_{ext} which is an external capacitor in parallel.

In Fig. 5.14a, the resonant frequency f_{ext} of the circuit is

$$f_{ext} = \frac{1}{2\pi\sqrt{L_C\left(C_p + C_{ext}\right)}} \tag{5.27}$$

After the parallel capacitor C_{ext1}, the resonance frequency f_{ext1} of the circuit in Fig. 5.14b becomes

$$f_{ext1} = \frac{1}{2\pi\sqrt{L_{ext1}\left(C_p + C_{ext1}\right)}} \tag{5.28}$$

In the formula, L_{ext1} refers to the inductance value of the inductor at the resonant frequency f_{ext1}. Similarly, if the value of the shunt capacitor is C_{ext2}, the resonant frequency f_{ext2} of the circuit becomes

$$f_{ext2} = \frac{1}{2\pi\sqrt{L_{ext2}\left(C_p + C_{ext2}\right)}} \tag{5.29}$$

where L_{ext2} is the inductance of the inductor at frequency f_{ext2}. Combined Eqs. (5.28) and (5.29):

$$\frac{f_{ext1}}{f_{ext2}} = \frac{\sqrt{L_{ext2}\left(C_p + C_{ext2}\right)}}{\sqrt{L_{ext1}\left(C_p + C_{ext1}\right)}} \tag{5.30}$$

Through mathematical deformation, there is:

$$C_{\mathrm{p}} = \frac{f_{\mathrm{ext2}}^2 L_{\mathrm{ext2}} C_{\mathrm{ext2}} - f_{\mathrm{ext1}}^2 L_{\mathrm{ext1}} C_{\mathrm{ext1}}}{f_{\mathrm{ext1}}^2 L_{\mathrm{ext1}} - f_{\mathrm{ext2}}^2 L_{\mathrm{ext2}}} \qquad (5.31)$$

In the formula, if the appropriate capacitance values C_{ext1} and C_{ext2} are selected such that the resonance frequencies f_{ext1} and f_{ext2} are close to each other, it is considered that $L_{\mathrm{ext1}} = L_{\mathrm{ext2}}$, and the input Eq. (5.30) is available.

$$C_{\mathrm{p}} = \frac{f_{\mathrm{ext2}}^2 C_{\mathrm{ext2}} - f_{\mathrm{ext1}}^2 C_{\mathrm{ext1}}}{f_{\mathrm{ext1}}^2 - f_{\mathrm{ext2}}^2} \qquad (5.32)$$

Therefore, improving the high-frequency characteristics of the inductor can be achieved in two ways: one is to obtain the maximum inductance value under the same volume; the other is to reduce the distributed capacitance C_{p} of the coil as much as possible.

(5) Chip Capacitor

Chip capacitors, i.e., mica and ceramic capacitors, have low series resistance and inductance, so they are high-frequency capacitors. If the wires are short, and the frequency is as high as 500 MHz, some surface mount capacitors are available in the GHz band. These capacitors are commonly used in radio frequency (RF) circuits for filtering, bypassing, coupling, timing and frequency discrimination, and decoupling in high-speed digital circuits. In addition to high-k ceramic capacitors, they are generally stable to time, temperature, and voltage performance. Figure 5.15 shows the actual chip capacitors.

Ceramic capacitors have been used in high-frequency circuits for nearly 100 years, and the original ceramic capacitors were "wafer capacitors." However, due to the rapid development of ceramic capacitor technology in recent decades, ceramic capacitors now have many different styles, shapes, and sizes, and they are the "main force" of high-frequency capacitors.

Mica has a low dielectric constant, so the mica capacitor is larger in size than the capacitance value. The tremendous advances in integrated ceramic capacitor technology and the low capacitance-to- volume ratio performance of mica capacitor terminals have replaced ceramic mica capacitors in many low-voltage, high-frequency

Fig. 5.15 Chip capacitor

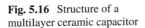

Fig. 5.16 Structure of a
multilayer ceramic capacitor

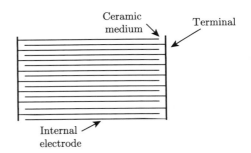

applications. Since mica has a high dielectric breakdown voltage, typically on the order of kilovolts, mica capacitors are still used in many high-voltage RF fields, such as radio transmitters.

Multilayer ceramic capacitors (MLCCS) consist of a multilayer ceramic material, usually a barium titanate medium, separated by crossed metal electrodes, as shown in Fig. 5.16. Put many capacitors in parallel. Some MLCCS have hundreds of layers of ceramic, each layer being only a few microns thick.

The advantage of this structure is that the capacitance of each layer can be doubled. The total capacitance is equal to the capacitance of each layer multiplied by the number of layers. At the same time, the inductance of each layer is divided. The total inductance is equal to the inductance of each layer times the number of layers. Multilayer capacitors combined with surface mount technology can produce nearly ideal high-frequency capacitors. Some small value (e.g., tens of pF) surface mount MLCCS may also have self-resonance over several gigahertz frequencies.

Most MLCCS have a capacitance value of 1 µF or less and a rated voltage of 50 V or less. The rated voltage is limited by a small layer gap. However, the combination of multiple layers of gaps allows manufacturers to fabricate MLCCS with large capacitance values of 10–100 µF. MLCCS are excellent high-frequency capacitors that are commonly used in high-frequency filtering and digital logic decoupling.

High-k ceramic capacitors are the only IF capacitors whose performance is unstable with respect to time, temperature, and frequency. Compared to standard ceramic capacitors, the main advantage of high-k ceramic capacitors is their large capacitance-to-capacitance ratio, which is commonly used for bypassing, coupling, and isolation of non-critical parts. Another disadvantage of high-k ceramic capacitors is that voltage transients can be damaged. Therefore, it is not recommended to bypass the bypass capacitor directly on the low-impedance power supply. Table 5.1 shows some typical capacitor fault scheme.

To suppress EMI noise on the PCB circuit, the most important thing is to make reasonable wiring during the circuit design and make the grounding system grounded well. However, in practical engineering applications, equipment manufacturers are often reluctant to redo circuits, considering cost and development time. At this time, it is necessary to consider chip capacitors.

Table 5.1 Typical capacitor failure scheme

Capacitor type	Usually used	Overvoltage
Aluminum electrolysis	Open road	Short circuit
Ceramic	Open road	Short circuit
Mica	Short circuit	Short circuit
Polyester film	Short circuit	Short circuit
Metallized polyester film	Leak	Noise
Solid tantalum	Short circuit	Short circuit

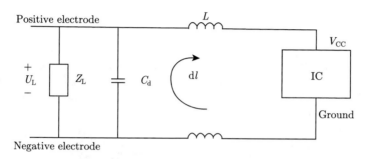

Fig. 5.17 Power supply decoupling network

In the electronic device PCB circuit, power supply decoupling of noise source devices such as digital ICS, power electronic devices, and relays is a convenient and effective noise suppression method.

As shown in Fig. 5.17, the power supply decoupling network, C_d is a decoupling chip capacitor, paralleled with the power supply terminal and ground terminal of the noise source device. The decoupling capacitor provides a charge source adjacent to the noise source device. When the noise source device switches, the decoupling capacitor can provide the required transient current through a low-impedance path.

In addition, the decoupling capacitor provides a low AC impedance between the power supply and the ground line, which effectively shortened the noise injected back into the power supply or grounded system by the noise source device. The capacitance of the decoupling capacitor is generally selected according to the principle of "the higher the frequency, the smaller the capacitance, the lower the frequency, the larger the capacitance," generally three orders of magnitude: 0.001, 0.01, and 0.1 µF.

2. Shortening the Distance of the Transmission Line

(1) Noise Model and Equivalent Circuit

There are various types of transmission cables in the PCB circuit, which are equipped with high-speed digital signals. If the length of the transmission cable is the same order of magnitude as the wavelength corresponding to the mounted high-frequency digital signals, the electric dipole radiation model is no longer applicable, consider the slender straight antenna model, as shown in Fig. 5.18.

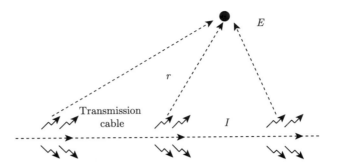

Fig. 5.18 Radiated EMI noise model caused by transmission cable

Without loss of generality, suppose the radiation antenna (transmission cable) is placed along the z-axis, and the current J on the surface of the radiation antenna is along the z-axis direction, so the vector potential A has only the z component, according to Lorentz conditions

$$\nabla \cdot A + \frac{1}{c^2} \frac{\partial \varphi}{\partial t} = 0 \quad E_x = -\frac{\partial \varphi}{\partial x} - \frac{\partial A_x}{\partial t} \tag{5.33}$$

where A is the vector potential, V s/m; A_z is the z component of the vector potential, V s/m; E_z is the z component of the electric field, V/m; φ is the potential, V; c is the speed of light, m/s.

From Eq. (5.33), there is:

$$\frac{1}{c^2} \frac{\partial E_x}{\partial t} = -\frac{1}{c^2} \frac{\partial}{\partial x} \left(\frac{\partial \varphi}{\partial t} \right) - \frac{1}{c^2} \frac{\partial^2 A_x}{\partial t^2} = \frac{\partial^2 A_x}{\partial x^2} - \frac{1}{c^2} \frac{\partial^2 A_x}{\partial t^2} \tag{5.34}$$

If the radiation antenna is an ideal conductor, the tangential component E_z of the electric field is zero on the surface of the radiation antenna, so the vector potential A_z meets the one-dimensional wave formula on the surface of the radiation antenna.

$$\frac{\partial^2 A_z}{\partial z^2} - \frac{1}{c^2} \frac{\partial^2 A_z}{\partial t^2} = 0 \tag{5.35}$$

It can be seen from Eq. (5.35) that along the surface of the radiation antenna, the vector potential $A_z(z)$ is a form of fluctuation, where vector potential A is

$$A(x) = \frac{\mu_0}{4\pi} \int \frac{J(x') e^{jkr}}{r} dU' \tag{5.36}$$

Let the length of the transmission cable be l, and the current distribution in the cable is

$$I(z) = \begin{cases} I_0 \sin k \left(\dfrac{l}{2} - z \right) & 0 \leqslant z \leqslant \dfrac{l}{2} \\[2mm] I_0 \sin k \left(\dfrac{l}{2} + z \right) & -\dfrac{l}{2} \leqslant z \leqslant 0 \end{cases} \tag{5.37}$$

where I and I_0 are the current distribution in the cable and the current at the center of the cable, respectively, A; k is a constant. Substituting Eq. (5.37) into Eq. (5.36) results in radiated electromagnetic fields B and E due to the transmission cable.

$$\begin{aligned} B &= i \frac{\mu_0 e^{ikR} I_0}{2\pi R \sin \theta} \left(\cos \frac{kl}{2} - \cos \frac{k(l-\lambda)}{2} \cos \frac{k\lambda \cos \theta}{2} \right. \\ &\quad \left. + \cos \theta \sin \frac{k(l-\lambda)}{2} \sin \frac{k\lambda \cos \theta}{2} \right) e_\phi \\ E &= j \frac{\mu_0 c e^{jkR} I_0}{2\pi R \sin \theta} \left(\cos \frac{kl}{2} - \cos \frac{k(l-\lambda)}{2} \cos \frac{k\lambda \cos \theta}{2} \right. \\ &\quad \left. + \cos \theta \sin \frac{k(l-\lambda)}{2} \sin \frac{k\lambda \cos \theta}{2} \right) e_\theta \end{aligned} \tag{5.38}$$

(2) Noise Suppression Method

There are generally long transmission cables between PCB boards, PCB circuit, and other systems. Noise currents generate strong radiated noise through the transmission cables. It can be seen from Eq. (5.38) that the radiated noise is related to cable length, noise current, and equivalent impedance. Radiated noise can therefore be reduced by shortening the distance of the transmission line.

3. Good Ground

(1) Noise Model and Equivalent Circuit

In high-speed digital printed circuits, due to poor grounding of various types of devices or the ground potential of the grounding point is not zero, it is easy to generate electric dipole radiation, that is, common mode radiation. According to Maxwell's formulas and electric dipole radiation theory, the common mode radiated field can be expressed as

$$\begin{cases} H_\phi = \dfrac{I dl k^2}{4\pi} \left[-\dfrac{1}{jkr} + \dfrac{1}{(kr)^2} \right] \sin \theta e^{-jkr} \\[3mm] E_\theta = \dfrac{I dl k^3}{4\pi \omega \varepsilon_0} \left[\dfrac{-1}{jkr} + \dfrac{1}{(kr)^2} + \dfrac{1}{(kr)^3} \right] \sin \theta e^{-jkr} \\[3mm] E_r = \dfrac{I dl k^3}{2\pi \omega \varepsilon_0} \left[\dfrac{1}{(kr)^2} + \dfrac{1}{j(kr)^3} \right] \cos \theta e^{-jkr} \end{cases} \tag{5.39}$$

where H is the magnetic field strength, A/m^2; E is the electric field strength, V/m; Idl is the electric moment of the electric dipole, A/m; k is the wave vector, rad/m, the mode represents the wave number, and the direction represents the direction of wave propagation; r is the test distance, m; ε_0 is the vacuum dielectric constant, F/m; ω is the angular frequency, rad/s.

It can be seen from Eqs. (5.39) and (4.52) that the near-field wave impedance of the common mode radiation is inversely proportional to the test distance and is high impedance (greater than $120\pi\Omega$). On the other hand, the common mode radiated field is related to the length of the equivalent short straight antenna, and the longer the equivalent length, the larger the common mode radiated field strength.

(2) Noise Suppression Method

The radiated noise caused by poor grounding can be equivalent to electric dipole radiation. The grounding fault in the PCB circuit is mainly caused by the ground potential of the grounding point being nonzero. For example, the high-frequency signal grounding system impedance caused by the wiring problem is too large, it is used to isolate the ferrite bead between the power sources at different levels, and the grounding potential may be different due to multi-point grounding. It can be seen from Eq. (5.39) that the radiated noise is related to the length of the equivalent short straight antenna and the ground potential of the grounding point. Therefore, the signal lines connecting the devices should not be too long, and the system ground should be strengthened.

For high-frequency signals, as the signal frequency increases, the transmission impedance of the signal line cannot be ignored. Therefore, the transmission impedance of the signal line should be calculated in the early stage of PCB design, and the appropriate wiring method should be designed to improve the grounding performance of the system. On the other hand, since the ferrite bead has a nonlinear frequency response characteristic and the impedance corresponding to the characteristic frequency is the largest, the ferrite bead between the power supply grounds of each stage is liable to cause poor system grounding. In addition, the use of multi-point grounding may result in different potentials at each grounding point. However, using a single-point grounding method may also increase the system grounding impedance. Therefore, using PCB multi-layer board design and ground surface (stratum) grounding can help to improve the grounding performance of the system.

As shown in Fig. 5.19, the 3 V DC power supply is connected to three CMOS inverters. The first inverter is triggered by a 25 MHz clock with both rise and fall times equal to 1 ns. Different power and signal conductor layouts are presented to illustrate how these different wiring approaches affect the grounding system.

Figure 5.20a shows a first layout arrangement using two parallel conductors 10 mm apart. Figure 5.20b shows the simulation results of the ground bounce, that is, the potential difference between the power supply 0 V reference voltage and the load ground. The ground bounce fluctuates between −0.816 and +2.11 V. Figure 5.21a shows another layout structure using microstrip lines. The top conductor is used for power routing, and the ground is used for ground loops. The distance between the top trace and the ground plane is 1.5 mm, which is much smaller than the first trace. Figure 5.21b shows the simulation results of the ground bounce. The ground bounce now swings between −0.334 and +0.8 V, which is significantly reduced compared to

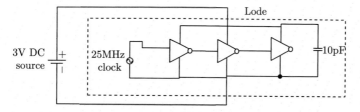

Fig. 5.19 Three inverters powered by a 3 V DC source

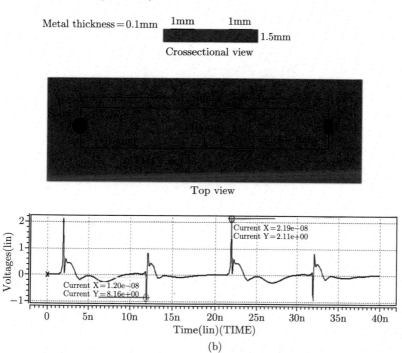

Fig. 5.20 Two parallel conductors with 10 mm separation

the first layout. This example further confirms that if two conductors (signal to ground or power to ground) are closely spaced, it affects system grounding performance.

In order to illustrate the quality of the ground plane design, simulation was carried out using electromagnetic simulation software. A simple microstrip line carrying a 200 MHz signal is used as an example of simulation. Figure 5.22 shows the ground loop for different ground plane configurations.

Figure 5.22a shows the current distribution at ground level. As expected, the signal return path mirrors the output signal conductor. This is the best layout practice because the current return path is undisturbed, the signal loop area is the smallest, and the ground bounce is also the lowest.

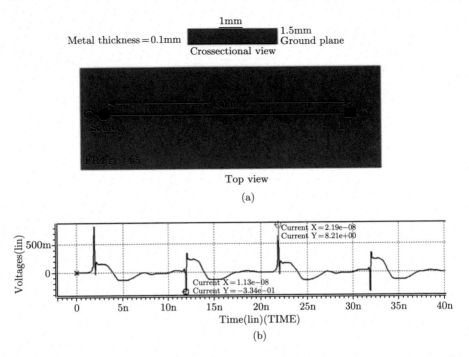

Fig. 5.21 Microstrip layout with 1.5 mm separation

Figure 5.22b shows the current distribution in the ground plane with a long side parallel to the slot of the output signal conductor. The loop current path is slightly disturbed and the loop area is larger than Fig. 5.22a. Therefore, it is expected that there is higher radiation and ground bounce.

Figure 5.22c shows the current distribution on the ground plane where the slots are directly orthogonal to the output signal conductors. The ground return current is severely interrupted, and the radiation and ground bounce are much higher than the first two cases.

Finally, Fig. 5.22d shows the worst practice of one side of the socket touching the edge of the PCB. The ground return current is forced to redistribute along the edge of the slot. In this case, the radiation and ground bounce are the highest.

5.2.4   Differential Mode Suppression Method

Different radiated EMI noise mechanisms correspond to different noise suppression methods, and usually because some parameters such as signal operating frequency and chip-rated driving current, which are selected according to functional requirements during circuit design, generally cannot be changed.

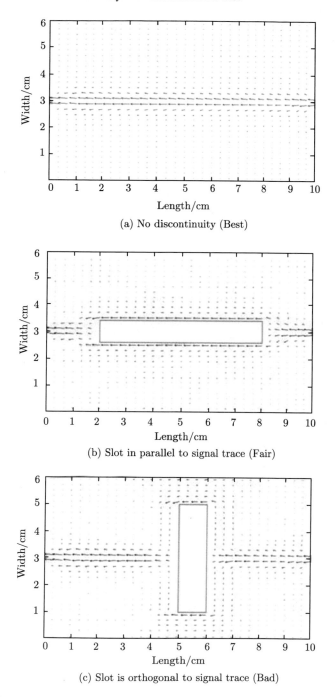

(a) No discontinuity (Best)

(b) Slot in parallel to signal trace (Fair)

(c) Slot is orthogonal to signal trace (Bad)

Fig. 5.22 Current return paths in ground plane

Fig. 5.22 (continued)

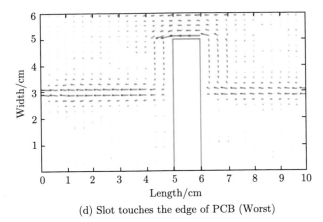

(d) Slot touches the edge of PCB (Worst)

The differential mode radiated field is proportional to the square of the signal frequency, the signal current loop area, and the differential mode current, and is inversely proportional to the test distance. The methods that can be taken according to the diagnosis result of the radiation mechanism mainly include weakening the differential mode radiated field by reducing the signal current loop area and reducing the I_{DM} (filtering).

1. Reduce Loop Area

(1) Noise Model and Equivalent Circuit

In the PCB, radiated noise caused by a large loop of the signal is often present, which can be equivalent to magnetic dipole radiation, that is, differential mode radiation. Similarly, according to Maxwell's formulas and magnetic dipole radiation theory, the differential mode radiation field can be expressed as Eq. (5.40).

$$\begin{cases} H_\theta = \dfrac{I d S k^3}{4\pi}\left[-\dfrac{1}{kr} - \dfrac{1}{j(kr)^2} + \dfrac{1}{(kr)^3}\right]\sin\theta e^{-jkr} \\[2mm] H_r = \dfrac{I d S k^3}{2\pi}\left[\dfrac{-1}{j(kr)^2} + \dfrac{1}{(kr)^3}\right]\cos\theta e^{-jkr} \\[2mm] E_\phi = \dfrac{I d S k^4}{4\pi\varepsilon_0\omega}\left[\dfrac{1}{kr} + \dfrac{1}{j(kr)^2}\right]\sin\theta e^{-jkr} \end{cases} \qquad (5.40)$$

In the formula, $I d S$ is the magnetic moment of the magnetic dipole, A/m^2.

It can be seen from Eqs. (5.41) and (4.55) that the near-field impedance of the differential mode radiation is proportional to the test distance and is low impedance (less than $120\pi\Omega$). On the other hand, the differential mode radiated field is related to the area of the equivalent loop antenna. The larger the equivalent area, the larger the differential mode radiated field strength.

(2) Noise Suppression Method

The radiated noise caused by the large loop of the signal can be equivalent to the magnetic dipole radiation, which is easy to generate serious radiation noise, such as signal line and ground loop caused by poor PCB wiring, signal loop for RFID communication, and loop radiation caused by excessive differential signal distance. It can be seen from Eq. (5.40) that the radiated noise is related to the noise current and the area of the equivalent loop antenna. Therefore, the signal line and the ground line in the circuit should avoid forming a large loop, and the distance between the differential signal lines should be reduced. In addition, for the communication signal loop, full capacitance filter can be used to reduce the noise current in the signal loop and improve the signal-to-noise ratio.

2. Reduce IDM (Filtering)

(1) Noise Model and its Equivalent Circuit

Due to the interference of the switching device, the RF antenna and various RF signals, many RF electromagnetic fields are coupled to the cable in the form of electromagnetic induction, generating a high-frequency noise current I_E. The high frequency noise current is easily superimposed with the effective signal I transmitted in the cable. After passing through the nonlinear device in the PCB circuit, mixing is likely to occur, causing distortion of the signal in the circuit, thereby generating a large amount of radiated EMI noise, as shown in Fig. 5.23. According to Faraday's law of electromagnetic induction, the induced current is

$$
\begin{aligned}
I_E &= I_{E1} + I_{E2} \\
I_{E1} &= \frac{1}{R} \int_L (\boldsymbol{v} \times \boldsymbol{B}) \cdot \mathrm{d}\boldsymbol{l} \\
I_{E2} &= -\frac{1}{R} \frac{\mathrm{d}\boldsymbol{\Phi}}{\mathrm{d}t}
\end{aligned}
\tag{5.41}
$$

In the formula, I_E is the cable-induced current caused by crosstalk, A; I_{E1} and I_{E2} are the motion current and the induced current, respectively, A; R is the cable equivalent resistance, Ω; v is the speed, m/s; $\boldsymbol{B}$ is the radiated magnetic field, T; L is the length of the cable, m; $\boldsymbol{\Phi}$ is the magnetic flux, Wb; t is the time, s.

Set the cable effective signal and the induced current to

$$
I = I_{m1} \cos(\omega_1 t + \varphi_1) \quad I_E = I_{m2} \cos(\omega_2 t + \varphi_2)
\tag{5.42}
$$

where, I_{m1} and I_{m2} are the amplitudes of the signals, respectively, A; ω_1 and ω_2 are the angular frequency, respectively, rad/s; φ_1 and φ_2 are the initial phase of the signal, respectively, rad/s.

The mixed current generated by the nonlinear device in the PCB is

$$
I' = I \times I_E = \frac{1}{2} A_1 A_2 \left[\cos((\omega_1 + \omega_2) t + \varphi_1 + \varphi_2) + \cos((\omega_1 - \omega_2) t + \varphi_1 - \varphi_2) \right]
\tag{5.43}
$$

Fig. 5.23 Radiated EMI
noise model caused by
crosstalk

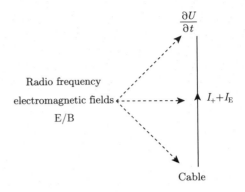

In the formula, I' is the noise current after mixing. Two groups of signals with different frequencies produce sum frequency (up conversion) and difference frequency (down conversion), which lead to signal distortion, thus produce electromagnetic interference, and even affect the normal operation of the system.

(2) Noise Suppression Method

Due to switching power supply, crystal oscillator, single chip microcomputer, DSP, and other high-speed signal processing chips, PCB circuit generates many radio frequency electromagnetic fields, which causes a lot of EMI noise to crosstalk to the signal line. After passing through the nonlinear components of the circuit, noise signal and working signal produce mixing and signal distortion, which affects the normal operation of electronic system, and is accompanied by strong EMI noise.

It can be seen from Eq. (5.45) that the crosstalk noise is related to the induced current. Therefore, the corresponding EMI filter can be designed at the signal line port to reduce the influence of crosstalk noise according to the operating signal frequency, bandwidth, and crosstalk noise frequency and bandwidth. However, the characteristic parameters of the EMI filter such as the passband, start frequency, cutoff frequency, and insertion loss are related to crosstalk noise suppression. In addition, improper design of the input and output impedance of the EMI filter may cause impedance mismatch of the signal line, causing signal reflection and other problems.

On the other hand, the electromagnetic shielding measures adopted by the signal lines are poorly applied, and the crosstalk noise is related to the shielding effectiveness of the material, especially the high-frequency noise above 800 MHz.

In the early stage of PCB design, the signal line can be changed to differential signal I_+, I_-, where the working signal I is

$$I = I_+ - I_- \tag{5.44}$$

In the formula, the differential signals I_+ and I_- have the same amplitude and opposite phases. After the crosstalk of the RF electromagnetic field, the differential signal becomes $I_+ + I_E$, $I_- + I_E$, as shown in Fig. 5.24, but the working signal is

$$I = (I_+ + I_E) - (I_- + I_E) = I_+ - I_- \tag{5.45}$$

Fig. 5.24 Crosstalk noise suppression mechanism using differential signals

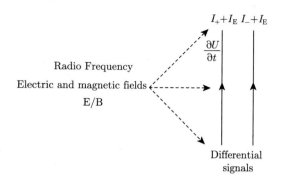

The working signal is not affected by the crosstalk of the RF electromagnetic field. At the same time, reducing the length of signal line as much as possible and using crosstalk choke coil can greatly reduce the radiated noise caused by crosstalk.

5.3 Case Study

Aiming at the radiated EMI noise generated by complex electronic system, Chap. 4 proposes an effective scheme to suppress the radiated EMI noise. This section discusses the excessive reasons for the radiated EMI noise combined with the products of power equipment and medical equipment, and designs an ideal scheme to suppress the radiated EMI noise, so that it can pass the GB 9254 standard detection and provide practical guidance for engineering application of EMI noise suppression scheme for complex electronic systems.

5.3.1 Case Study 1

1. Product Introduction and the Description of Problem

Figure 5.25 presents the overall structure of the sodium–sulfur energy storage converter. The object of this study is shown in Fig. 5.26. This module is the communication module of the sodium–sulfur energy storage converter. Because of its small size, complex structure, and high integration of digital circuits, the communication module is processed separately.

The radiation test standard is Class B, and the test range is 30 MHz–1 GHz. The radiated EMI noise of communication module of sodium–sulfur energy storage converter was tested in Jiangsu Electrical Equipment EMC Engineering Laboratory in 3M anechoic chamber and EMI receiver (LP3) made by German company R&S.

Preliminary EMI results of the communication module of sodium–sulfur energy storage converter for batteries are shown in Fig. 5.27. Obviously, the peak value

Fig. 5.25 Overall structure of sodium–sulfur energy storage converter

the sodium-sulfur energy storage converter

Fig. 5.26 Actual wiring diagram of control module

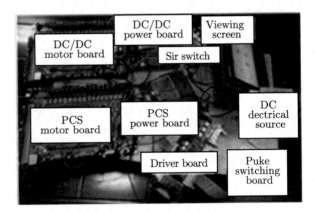

near the 40 MHz frequency band has approached the critical line, and there is no appropriate margin left. The radiated noise of 60–250 MHz exceeds the standard seriously. The test results show that the radiated noise of the equipment has not passed the test.

2. EMI Inhibition Theory Analysis

The circuit block diagram of the control module is shown in Fig. 5.28, including DC/DC main board, PCS main board, DC/DC power board, PCS power board, relay, pulse converter board, IGBT driving board (sampling board), display screen, etc. The power converter converts 220 V AC to 24 V DC and supplies power to DC/DC power board and PCS power board. Both power boards output 15 V and 5 V DC, respectively. They supply power to DC/DC main board and PCS main board, respectively, and drive pulse converter board, the signal voltage amplitude is 3.3 V. The signal modulated by the pulse converter board is sent to the sampling board to control the working state of the inverter, and the voltage amplitude of the signal is 15 V. There is a CAN communication mode connection between PCS and DC/DC main board. PCS main board is connected with display screen to display data.

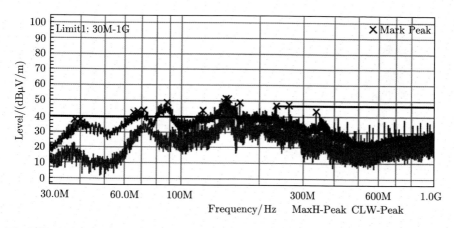

Fig. 5.27 Control module radiation interference preliminary test results (failed)

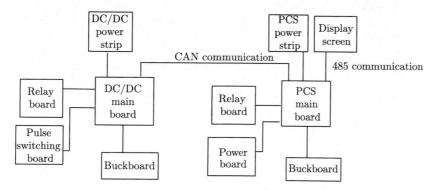

Fig. 5.28 Circuit block diagram of control module for sodium–sulfur energy storage converter

3. Analysis of EMI Rectification Methods

The electric field probe is used to detect the chips on PCB board. The signal entrance of PCS board and DC/DC board, the DSP chip of PCS board and DC/DC board, PCS board and DC/DC board are noisy. The main board DSP chips have obvious noise at 54 MHz and 172 MHz, and the signal entry of the main board is similar at 53 MHz, as shown in Fig. 5.29.

The radiated noise in Fig. 5.27 is concentrated in the frequency band of 30–150 MHz. According to the division of frequency band, it includes the low-frequency radiated noise band and the front part of the medium frequency radiated noise band. According to the design of low-frequency radiated filter, capacitor and inductance are selected to filter, and ferrite core inductor filter is considered.

The decoupling capacitor is selected according to the preliminary test results, the capacitance value is inversely proportional to the noise over-punctuation frequency. Generally, three orders of magnitude are selected, namely 0.001, 0.01, and 0.1 μF. For example, in order to suppress the high-frequency noise of this product, the 0.001 μF

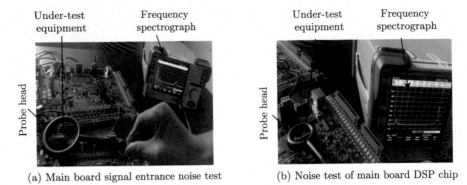

(a) Main board signal entrance noise test   (b) Noise test of main board DSP chip

Fig. 5.29   Measurement of high-frequency noise with electric field probe

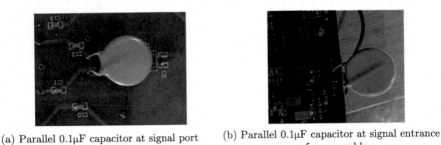

(a) Parallel 0.1μF capacitor at signal port   (b) Parallel 0.1μF capacitor at signal entrance
                                                                    of power cable

Fig. 5.30   Control module radiation modification

capacitor is selected first, but the effect is not obvious. At this time, the 0.01 μF
capacitor is used instead, if the noise is degraded but still not ideal enough to switch
to a higher order of capacitor. After several modifications, there is negative feedback.
At this time, several lower-order capacitors can be selected to stack between the two
orders to achieve better results. According to repeated attempts and amendments,
the ceramic capacitor filter with 0.1 μF at both ends of the main board power cable
signal entry and the main board DSP signal port is finally achieved, as shown in
Fig. 5.30. Since capacitors have successfully modified the radiation, the use of ferrite
core inductor is no longer considered.

4. Final Result After Modification

Based on the modification of Sect. 5.3.1'(3)', the radiation interference test of the
control module is carried out again, the test results are shown in Fig. 5.31. The
capacitive filter is used to suppress low-frequency radiated noise, and the problem of
electromagnetic interference is solved at a very small cost. Compared with the initial
measurement of radiation interference in Fig. 5.27, it can be seen that the radiated
noise has a good suppression effect, and the noise of 60–250 MHz has been reduced
to below the standard line.

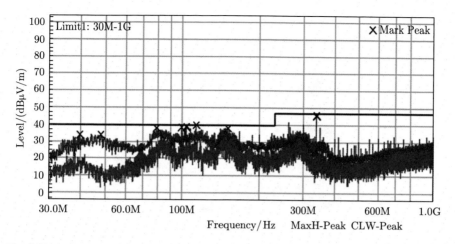

Fig. 5.31 Control module radiation interference correction result diagram (passed)

Table 5.2 Comparison between radiation preliminary test and modification

Test frequency (MHz)	39.9425	71.2250	87.5695	145.8180	166.6245
Initial value (dB μV/m)	38.11	44.09	49.20	49.46	48.91
Rectification result (dB μV/m)	33.90	29.08	45.22	38.24	34.28
Inhibition (dB μV/m)	4.21	7.85	15.01	11.22	14.63
Safety margin (dB μV/m)	6.10	10.92	4.78	11.76	15.72

The representative peaks and noise values in the preliminary measurement of radiation interference (Fig. 5.27) and the correction results (Fig. 5.31) are sorted out, and Table 5.2 is obtained. The maximum noise suppression is 15.01 dB μV/m, and the maximum peak value of the amplitude remains a 4.78 dB μV/m safety margin, which meets the test standard.

5.3.2 Case Study 2

1. Product Introduction and the Description of Problem

The appearance of meridian leveling therapeutic instrument is shown in Fig. 5.32. In the course of treatment, special low frequency uninterrupted pulses can be produced,

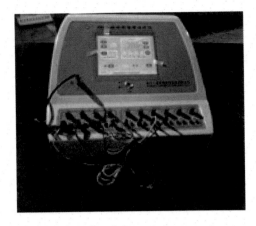

Fig. 5.32 Appearance of meridian leveling therapeutic instrument

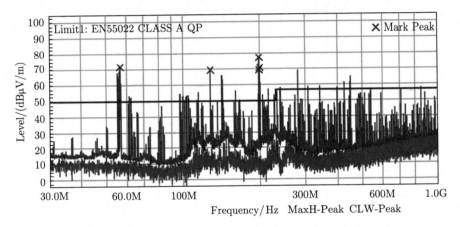

Fig. 5.33 Preliminary test of meridian leveling therapeutic instrument (failed)

and the ultrahigh output pulse voltage of 0–3000 V pulse can be adjusted after power amplification. In this process, high-frequency noise can easily be generated.

According to standard, the equipment is tested at level A. The preliminary measurement of radiated electromagnetic interference noise of meridian instrument shows that the radiated interference is almost in full frequency band after 50 MHz as shown in Fig. 5.33. The radiated noise at the frequency points of 56.578 MHz and 198.2465 MHz is 72.00 dB μV/m and 77.40 dB μV/m, respectively, which exceed the radiation test A standard of 22.00 dB μV/m and 7.40 dB μV/m, respectively. Therefore, the meridian instrument has not passed the standard test, so it needs to take measures to suppress the electromagnetic interference noise, so that the radiated noise is below the standard limit, and there is a certain margin.

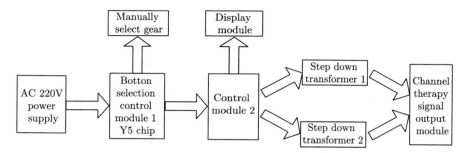

Fig. 5.34 Function module of meridian leveling therapeutic instrument

2. EMI Inhibition Theory Analysis

Figure 5.34 is the functional module diagram of meridian leveling therapeutic instrument. After connecting 220 V AC, signal is selected manually through control module 1 (Y5 chip). Control module 2 chooses step-down transformer 1 and step-up transformer 2 according to manual selection, which generates ultrahigh output pulse voltage of 0–3000 V pulse. At the same time, display module shows the corresponding voltage level.

3. Analysis of EMI Rectification Methods

Analysis of the preliminary test of Fig. 5.33 meridian leveling therapeutic instrument shows that the radiated noise of meridian instrument is concentrated in 50 MHz–1 GHz, including almost all the frequency bands of the cable integrated hybrid filter. Considering the rectification from the high-frequency band, according to the shielding principle proposed in the design of high frequency radiation filter, the transformer is considered, adding metal bushing and choosing sticky copper foil in metal bushing to wrap transformer is convenient to install and not easy to fall off.

In the process of practical engineering renovation, high-frequency devices, transformers, and so on are common noise sources. These devices are disconnected in advance. If the radiated noise of the system is significantly reduced, the device is initially identified as the main noise source. As shown in Fig. 5.35, after disconnecting the isolation transformer in the meridian instrument, the radiation emission test is carried out again, and Fig. 5.36 and Table 5.3 are obtained. Obviously, after the disconnection of the isolation transformer, the radiation test value of the whole meridian leveling therapeutic instrument decreases, so it is inferred that the isolation transformer is the noise source, which can be rectified.

Modification (1): Fig. 5.37 adds ferrite core inductor to the input of transformer and wraps copper foil.

The test results of Fig. 5.38 show that the transformer modification measures have obvious effect, the copper foil shielding ability is remarkable, and the over-standard noise after 200 MHz is completely suppressed, especially the 283.4417 MHz frequency points are reduced by 28.75 dB μV/m, and the 449.2378 MHz frequency points are reduced by 34.28 dB μV/m. It presents that the over-standard still exists and needs further rectification.

Disconnect the isolation transformer
to determine the noise source

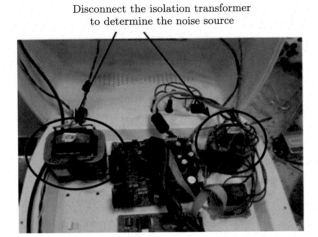

Fig. 5.35 Disconnecting isolation transformer

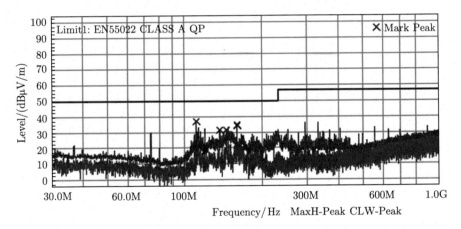

Fig. 5.36 Radiation test result of disconnected isolation transformer

Table 5.3 Comparisons of radiation result before and after transformer modification

Test frequency (MHz)	56.5780	127.8735	198.2465	283.4417	449.2378
Initial value (dB μV/m)	72.00	69.86	77.40	69.78	65.82
Rectification result (dB μV/m)	50.01	54.56	52.55	41.03	31.54
Inhibition (dB μV/m)	11.99	15.30	24.85	28.75	34.28
Safety margin (dB μV/m)	−0.01	−4.45	−2.55	−8.97	18.46

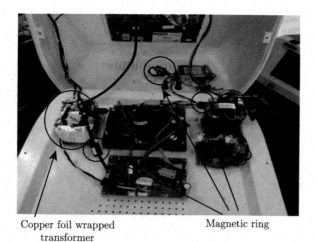

Copper foil wrapped   Magnetic ring
transformer

Fig. 5.37   Transformer rectification measures

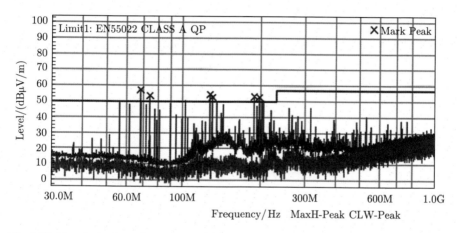

Fig. 5.38   Test result of transformer modification

Modification (2): As shown in Fig. 5.39, according to the test results obtained by Modification (1), the radiated noise has been suppressed after the change from full-band over-standard to 300 M. This shows that the suppression of copper foil is successful. Obviously in Fig. 5.38, the noise becomes a spike shape with a certain distance, and it can be judged that the radiation source in PCB board is mainly the signal end of MCU. Therefore, 22 pF patch capacitors are connected in parallel on SCLK, RXD1, TXD2, SDA, STR, RXD1, TXD1, SCK, and RST pins, respectively, to form parallel resonant filter in the circuit.

When many of the same $L-C$ networks are connected in parallel, C is the capacitance of a network and n is the number of parallel networks. The total capacitance C_t is [6]

Fig. 5.39 Pin parallel capacitance of single-chip microcomputer

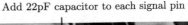

Add 22pF capacitor to each signal pin

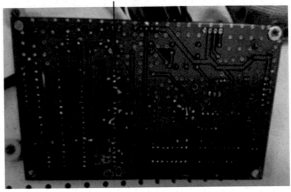

$$C_t = nC \tag{5.46}$$

The total inductance of N parallels $L-C$ networks is L_t.

$$L_t = \frac{L}{n} \tag{5.47}$$

L is the inductance of a network. When two conductors are connected in parallel, the mutual inductance effect must be considered in calculating the total inductance. The net inductance of two parallel conductors with the same current can be written as [6]

$$L_t = \frac{L_1 L_2 - M^2}{L_1 + L_2 - 2M} \tag{5.48}$$

L_1 and L_2 are part self-inductance of two conductors, and M is part mutual inductance between them. If two conductors are identical (L_1 equals L_2), Eq. (5.48) is simplified to

$$L_t = \frac{L_1 + M}{2} \tag{5.49}$$

Equation (5.49) shows that mutual inductance limits the reduction of total inductance of shunt conductors. From Eqs. (5.47) and (5.48), when the same $L-C$ network is parallel, capacitance (good parameters) is multiplied by the number of networks, while inductance (bad parameters) is divided by the number of networks. Therefore, the inductance of the network can be reduced to any desired value by using enough parallel $L-C$ networks. According to this theory, 22 pF capacitors are added to the signal pins of Fig. 5.37, respectively.

According to the results of Fig. 5.40 and the dotting records, Table 5.4 shows that the noise in the whole frequency band is obviously suppressed. Except for a peak close to the standard line at 150 MHz, there are many safety margins in the

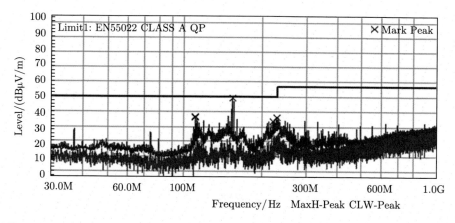

Fig. 5.40 Test result of single-chip microcomputer pin parallel capacitor

Table 5.4 Comparison before and after the signal pin of PCB single-chip microcomputer parallel with capacitor

Test frequency (MHz)	68.1212	47.1354	127.4366	188.3048	196.8420
Initial value (dB µV/m)	57.32	53.48	54.56	53.12	52.55
Rectification result (dB µV/m)	18.53	26.75	30.22	19.61	23.46
Inhibition (dB µV/m)	38.79	26.73	24.34	33.96	29.09
Safety margin (dB µV/m)	31.47	23.25	19.78	30.39	26.54

other frequency bands. The electromagnetic interference performance of meridian instrument is greatly improved, especially at 68.1212 MHz, the decrease value is about 38.79 dB µV/m. The decrease value at 188.3048 MHz frequency point is about 33.96 dB µV/m.

Modification (3): As shown in Fig. 5.41, the main chip pulse plus 200 pF capacitor along the road. After the Modification (2) rectification, most of the noise has been suppressed, and there is only a peak at 150 MHz, which indicates that this frequency point is the crystal oscillation frequency when the main chip of the single chip computer works. Noise frequency that needs to be suppressed is one of the most important considerations in choosing capacitors. Therefore, the 200 pF patch capacitor is chosen, which is parallel to the patch capacitor and connected to the main chip. Because of the small size of the patch capacitor and the absence of conductors, it is not easy to cause new coupling noise, and the parasitic inductance of the capacitor with conductors is significantly reduced. Therefore, the patch capacitor is a more effective high- frequency filter.

After the modification (3), the test results as shown in Fig. 5.42, 150 MHz frequency point has been suppressed, 153.4325 MHz frequency point, the noise decreased to 21.67 dB µV/m, and no other interference occurred, so the modification is completed. Table 5.5 is generated by comparing the data tested before and after the modification (3).

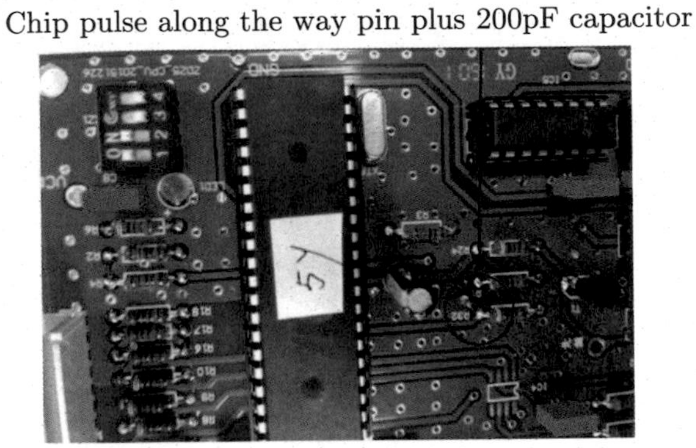

Chip pulse along the way pin plus 200pF capacitor

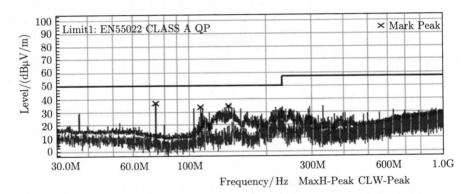

Fig. 5.41 Pulse chip modification

Fig. 5.42 Test result of 200 pF capacitance radiation along pulse circuit

Table 5.5 Comparisons of before and after adding 200 pF capacitor

Test frequency (MHz)	110.6070	150.2351	153.4325	153.4810	221.3325
Initial value (dB μV/m)	36.31	38.14	49.20	48.87	33.49
Rectification result (dB μV/m)	33.58	29.87	27.53	26.89	31.57
Inhibition (dB μV/m)	2.73	8.27	21.67	21.98	1.92
Safety margin (dB μV/m)	6.42	10.13	12.47	13.11	8.43

4. Final Result After Modification

After optimizing the wiring, the meridian leveling therapeutic instrument passed the GB 9254 Class A standard in the 3 M anechoic chamber of Jiangsu Electrical Equipment EMC Engineering Laboratory.

It shows that the initial radiation measurements of Fig. 5.33 and the radiation modification results of Fig. 5.42 are obviously suppressed and have a large safety margin in the over-standard frequency band near 50–250 MHz, which fully meets

Table 5.6 Comparison between original test and modification results

Test frequency (MHz)	56.5780	127.8735	198.2465	283.4417	449.2378
Initial value (dB μV/m)	72.00	69.86	77.40	69.78	65.82
Rectification result (dB μV/m)	17.36	32.41	21.88	25.93	22.15
Inhibition (dB μV/m)	54.64	37.45	55.52	43.85	43.67
Safety margin (dB μV/m)	32.64	17.59	28.12	31.07	34.85

the test standards. The data of Figs. 5.33 and 5.42 are collected and summarized to form Table 5.6. The final peak with the highest amplitude is left with a security margin of 43 dB μV/m.

References

1. Paul CR. A comparison of the contributions of common-mode and differential-mode currents in radiated emissions. IEEE Trans Electromagn Compat. 2002;31(2):189–93.
2. Hubing TH, Kaufman JF. Modeling the electromagnetic radiation from electrically small table-top products. IEEE Trans Electromagn Compat. 1989;31(1):74–84.
3. Hockanson DM, Drewniak JL. Investigation of fundamental EMI source mechanisms driving common-mode radiation from printed circuit boards with attached cables. IEEE Trans Electromagn Compat. 1996;38(4):557–66.
4. Zhao Y, See KY. A practical approach to education of electromagnetic compatibility at the undergraduate level. In: IEEE antennas and propagation society international symposium, vol. 6. Columbus, OH; 2003. p. 454–7
5. Zhao Y, See KY. A practical approach to EMC education at the undergraduate level. IEEE Trans Educ. 2004;47(4):425–9.
6. See KY, Manish O, Liu ZH, et al. Correlation between ground bounce and radiated emission. In: Electronics packaging technology conference, vol. 12. IEEE Xplore; 2004. p. 640–2.

Chapter 6
Principle and Analysis of EMS: Static Electricity Mechanism and Protection

6.1 ESD Generation Mechanism

Static electricity is present in our lives and it is a common phenomenon in daily life. The essence of static electricity is a phenomenon that occurs in the electrical energy of the surface of an object due to the local imbalance of positive and negative charges. Electrostatic phenomenon is a general term for the phenomenon of charge generation and disappearance. On the one hand, static electricity has many applications in our lives, such as electrostatic dusting, electrostatic spraying, electrostatic separation, and electrostatic copying. However, on the other hand, ESD can also cause certain damage to electronic products, circuits, and equipment, causing damage or instability of the functions.

This section discusses the mechanism of ESD, implementation standards, and test methods.

Electrostatic discharge (ESD) is an electrostatic charge transfer between objects caused by direct contact or electrostatic field induction. After the energy of the electrostatic field reaches a certain level, the discharging through the medium is called electrostatic discharge.

There are four types of electrostatic discharge (ESD) [1–3]:

(1) Frictional electricity generation: Frictional electrification is essentially the phenomenon of unbalanced positive and negative electric power after contact and separation. Friction is a process of constant contact and separation, which is the most common method of generating static electricity. The better the insulation of the material, the easier it is to rub the electricity.

© Science Press 2021
Y. Zhao et al., *Electromagnetic Compatibility*,
https://doi.org/10.1007/978-981-16-6452-6_6

(2) Contact separation generates electricity: Static electricity can be generated when two objects of different materials are separated after contact. When two different objects come into contact with each other, one loses the charge. For example, electrons are transferred to the other to make it positively charged, while the other gains electrons to make it negatively charged. If the charge is difficult to neutralize during separation, the charge may accumulate and make the object electrostatic.

(3) Electricity generation by induction: for a conductive body, since electrons can flow freely on the surface, if it is placed in an electric field, positive and negative ions transfer due to the mutual repulsion of like and mutual attraction of opposite.

(4) Conduction generation: for an electric conductor, electrons flow freely on the surface thereof. For example, in contact with a charged object, charge transfer occurs.

According to the principle of charge neutralization, when the static power source is in contact with other objects, the charge flow transfers enough power to offset the voltage. During the transmission of this high-speed power, potential voltages, currents, and electromagnetic fields are generated. In severe cases, the object to be contacted is destroyed, which is called ESD. The national standard is defined as follows: "ESD is the transfer of charge caused by objects with different electrostatic potentials in close proximity or direct contact." ESD can cause serious damage or malfunction of electronic equipment. Semiconductor experts and equipment users are looking for ways to suppress ESD.

In the production and use of electronic and electrical products, the operator is the most active static power source. When people wear fabrics of insulating materials and the shoes are insulated from the ground, people may accumulate a certain amount of electric charge when moving on the ground, and ESD occurs when the human body touches components or devices connected to the ground [1, 5].

When the distance from the ESD location is very close, both the electric field and the magnetic field are very strong, so the circuit near the ESD location is affected. Figure 6.1 shows the electromagnetic fields generated by a 4 kV ESD at different positions. It can be seen from Fig. 6.1 that the electromagnetic field generated by ESD not only has a short rise time, but also has a steep rising edge.

Therefore, most semiconductor devices are susceptible to damage from ESD, especially large-scale integrated circuit devices. The vulnerability of semiconductor devices is usually expressed by the value of the electrostatic voltage at which the pins and the insulating layer are broken by ESD [6–8]. Most semiconductor devices have a damage range of 100–300 V. Table 6.1 lists the range of reference values for common devices that are susceptible to damage.

The damage caused by static electricity to the device is divided into two types: dominant and recessive. Implicit damage is not easily detected, but the device becomes more fragile and can be easily damaged under conditions such as overpressure and high temperature.

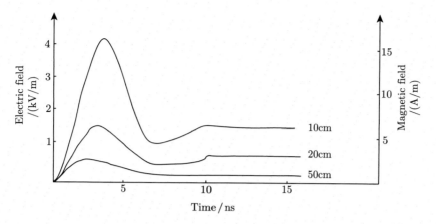

Fig. 6.1 Electromagnetic field generated by ESD

Table 6.1 Reference values for common devices that are susceptible to damage

Device type	Voltage range/V
Schottky diodes	300–2500
Schottky TTL	1000–2500
Bipolar transistor	380–7000
ECL	500–1500
Thyristor	680–1000
JFET	140–7000
CMOSFET	100–200
CMOS	250–3000
GaAsFET	100–300
EPROM	100

6.2 Electrostatic Protection Theory

6.2.1 Implementation Standards and Test Methods of ESD

The standards for ESD testing are GB/T 17626.2 and IEC 61000-4-2, which have different requirements for electromagnetic compatibility of electrical and electronic equipment under different environmental conditions. The test standard describes the instrument performance verification method, as well as the configuration [9, 10].

The IEC 61000-4-2 standard classifies ESD into different severity levels. The level of ESD is shown in Table 6.2. The 'X' is an undetermined level, which is determined by the equipment manufacturer and the user according to the actual situation.

Figure 6.2 shows a schematic diagram of the ESD generator. The pulse waveform when the tester is discharged does not depend on the high voltage power, but mainly

Table 6.2 Severity level of ESD

Level	Contact discharge test voltage (kV)	Air gap discharge test voltage (kV)
1	2	2
2	4	4
3	6	8
4	8	15
X	To be determined	To be determined

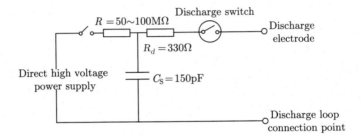

Fig. 6.2 Circuit of the ESD tester

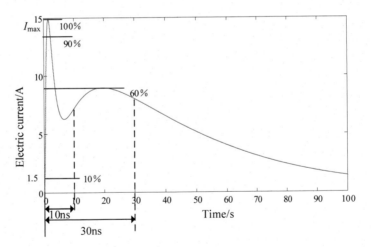

Fig. 6.3 Discharge current waveform of the ESD generator

depends on the type of storage capacitor, discharge resistor, and external load in the arrester. Figure 6.3 shows the discharge current waveform of the ESD generator, and I_m represents the current peak value. The rise time $t_r = (0.7 - 1)$ ns.

The main performance requirements of the IEC 61000-4-2 standard for ESD generators are given in Table 6.3. In the table, C_d is the distributed capacitance (which presents between the generator and the device under test or between the reference

Table 6.3 Main performance requirements of the ESD tester

Storage capacitor ($C_S + C_d$)	150 pF $\pm$ 10%
Discharge resistor (R_d)	330 $\Omega \pm 10\%$
Charging resistor (R_c)	50–100 MΩ
Output voltage: 8 kV (rating)	For contact discharge
Output voltage: 15 kV (rating)	For air gap discharge
Deviation of output voltage indication	$\pm 5\%$
Polarity of output voltage	Positive or negative
Duration	At least 5 s
Discharge method: single time	At least 1 s

ground plate and the coupling plate). The open circuit voltage measured on the storage capacitor is the output voltage. The test instrument generally has a discharge rate that at least 20 times per second.

A reference grounding plate made of copper or aluminum with a minimum thickness of 0.25 mm is required on the laboratory floor. The minimum area is 1 m². However, the actual size depends on the external dimensions of the equipment under test. It must be at least 0.5 m on each side of the equipment to be tested or the horizontal coupling plate on the test bench, and it must be connected to the earthing system. The equipment under test shall be connected according to the working condition, and at least 1 m away from the wall of the laboratory and other metal objects. The equipment under test is connected to the earthing system according to the installation regulations. No additional grounding is allowed. The power cord and signal cable should also be set according to the actual installation position [11–13].

The tester discharge return cable should be connected to the reference ground. The cable length is usually 2 m. If this length exceeds the length required for the selected discharge point, the excess should not be too close to the other conductive portions of the test configuration (at least 0.2 m), and there is no induction between the reference ground. The connection between the ground cable and the reference ground plane, as well as all solder joints, should be low impedance. The coupling plate used for indirect discharge should have the same material and thickness as the grounding plate, and be connected to the reference grounding plate through a cable with a 470 kΩ resistor on each end. These resistors can withstand the discharge voltage of the cable. It does not cause a short circuit when it is on the grounding plate.

Figure 6.4 shows the test environment of the desktop test equipment. Except as illustrated in the figure, the test equipment and the coupling plate are separated by 0.5 mm thick insulation. The horizontal coupling plate should be at least 0.1 m larger than the tested equipment. If the equipment to be tested is too large, a larger test bench can be used or two identical test stands can be used at the same time. A metal plate shall be placed at the joint and pressed against each table by more than 0.3 m.

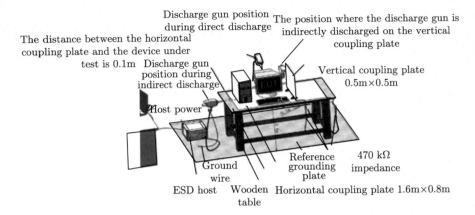

Discharge gun position The position where the discharge gun is
during direct discharge indirectly discharged on the vertical
The distance between the horizontal   coupling plate
coupling plate and the device under
test is 0.1m Discharge gun
           position during\   Vertical coupling plate
           indirect discharge\   0.5m×0.5m

Host power

                                              470 kΩ
                        Reference   impedance
              Ground   grounding
              wire   plate
       ESD host   Wooden Horizontal coupling plate 1.6m×0.8m
                   table

Fig. 6.4 Test environment for desktop devices

Fig. 6.5 Photograph of the
laboratory discharge test site

It is also required that two test stands are, respectively, connected to the reference ground by a cable with a resistance.

Figure 6.5 shows the ESD test site of an electronic product. The test uses contact discharge measurement. The ESD equipment used is the MODEL ESD-20 ESD meter of Suzhou TEST Electronic Technology Co., Ltd.

The two main disruption mechanisms of ESD are: thermal failure of equipment due to heat generated by ESD current, high voltage induced by ESD leads to insulation breakdown. Both types of damage can occur simultaneously in a single device. For example, a breakdown of insulation can trigger a large current, which leads to thermal failure of the equipment.

In addition to causing circuit damage, ESD can easily cause interference with electronic circuits. There are two ways in which ESD can interfere with electronic circuits. One way is to conduct, and the ESD current flows directly through the circuit, causing damage to the circuit. If a certain part of the circuit constitutes a discharge circuit, the ESD current directly intrudes into the device, such as manually touching the track, pins, I/O interface terminals of the device, and core wires of the coaxial socket. The ESD current flows through the input of the integrated chip, causing interference. The other way is radiation interference. A sharp current is generated during ESD. This current contains a high- frequency component, which generates a radiated magnetic field and an electric field. The magnetic field can induce an interference electromotive force in each signal loop of a nearby circuit. A large current change occurs in a short time (the aforementioned current change of 20 A in 150 ns), so the interference electromotive force generated in the signal loop may exceed the threshold level of the logic circuit, causing false triggering. The magnitude of the radiated interference also depends on the distance between the circuit and the ESD point, and the magnetic field generated by the ESD decays with the square of the distance. The electric field generated by ESD can be received by the track line on the PCB or the I/O line of the equipment, thus causing interference. The instantaneous peak of the electric field is very high, ranging from a few hundred kilovolts per meter to a few thousand kilovolts per meter, but decays with the cubic distance. When the distance is close, both the electric and magnetic fields are strong, indicating that the circuit near the ESD position is affected. Electromagnetic noise can enter electronic devices and circuits through conduction or radiation.

The basic way of radiation coupling can be capacitive or inductive, depending on the impedance of the ESD source and receiver. In a high-impedance circuit, the current signal is small, and the signal is represented by a voltage level. At this time, capacitive coupling is dominant, and ESD induced voltage is a major problem. In low-impedance circuits, the signal is in the form of a main current. Therefore, inductive coupling is dominant, and ESD-induced current causes problems in most circuits. In the far field, there is electromagnetic field coupling.

The upper limit of the electromagnetic interference energy associated with ESD can exceed 1 GHz, depending on the level, relative humidity, proximity speed, and shape of the discharged object. At this frequency, the typical equipment cable and even the traces on the plate become effective receiving antenna. Therefore, for typical analog or digital electronic devices, ESD can induce high-level noise.

Damage to the equipment is much more than the voltage and current required to cause malfunction. Damage is more likely to occur during conduction coupling. This means that if damage is caused, the ESD spark must be in directly contact with the circuit, and radiation coupling usually only causes malfunction.

Under the action of ESD, the devices in the circuit are more susceptible to damage under energized conditions than non-energized conditions.

6.2.2 ESD Suppression Device

1. TVS

Transient voltage suppresser (TVS) is an efficient circuit protection device. The response time of TVS is very fast and sub-nanosecond, so there is strong capability of static electricity protection [14]. TVS is a voltage regulator based on Zener diode. It is made of semiconductor material silicon (Si) and silicon carbide (SiC). The

Fig. 6.6 Physical structure
of TVS

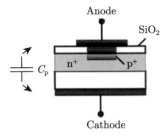

Fig. 6.7 Volt–ampere
characteristics of TVS

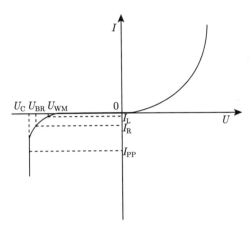

physical structure is shown in Fig. 6.6. The PN junction is composed of heavy-mixed III/V elements on both sides.

The volt-ampere characteristics of TVS are shown in Fig. 6.7. It has bidirectional voltage stabilization and strong current conductivity. Once the voltage at both sides of TVS exceeds the clamping voltage of TVS, TVS is instantaneously broken down, transforming from high impedance to low impedance and absorbing a large instantaneous current; thus, the voltage at both sides of the TVS is clamped near the identification value of device. Therefore, TVS has a good electrostatic protection effect on circuits. By paralleling the TVS between the power supply and the ground of the chip, the impact of static electricity on the chip can be avoided and the chip can work normally.

The equivalent circuit of TVS is shown in Fig. 6.8. Simplifying the distribution impedance, the equivalent circuit diagram of TVS suppressing static electricity is shown in Fig. 6.9. R_g is the internal resistance of power supply. TVS is connected in parallel at both sides of power supply. The relationship is $R_g > R_S + R_L > R_S$ between the internal resistance of power supply, and the distribution impedance of TVS static suppressor and the load satisfies.

The analysis of the circuit shows that the voltage at both sides of the load is [3]:

$$U_L = U_{BR} + \frac{R_S U_g}{R_g} \tag{6.1}$$

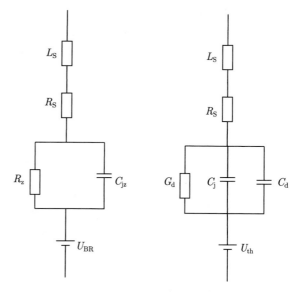

Fig. 6.8 Equivalent circuit of TVS under positive/negative voltage

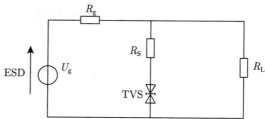

Fig. 6.9 Equivalent circuit of TVS on PCB board

2. Varistor

Zinc oxide varistor ceramic is a polycrystalline semiconductor ceramic element made by electronic ceramic technology with zinc oxide as the main body, added with a variety of metal oxides, and has nonlinear conductivity characteristics. It is the main component material for suppressing overvoltage, absorbing surge energy, and ESD protection.

The microstructure of zinc oxide varistor ceramics is shown in Fig. 6.10. It is composed of zinc oxide crystal grains and grain boundary materials. The zinc oxide crystal grains are doped with donor impurities to form an N-type semiconductor, and the grain boundary material contains a large amount of metal oxides to form a large number of interface states. Therefore, two crystal grains and a grain boundary (i.e., microscopic unit) can form a back-to-back bidirectional NPN junction, and the entire ceramic is composed of many back-to-back bidirectional NPN junctions in series and parallel [15].

Because zinc oxide varistor ceramics have very thin grain boundaries (1 nm or less), when the applied voltage is less than the breakdown voltage of the reverse PN junction, it belongs to the Schottky barrier thermionic emission conductance,

Fig. 6.10 Microstructure

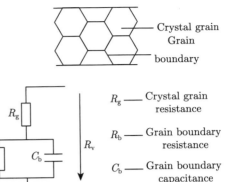

— Crystal grain
  Grain
  boundary

Fig. 6.11 Equivalent circuit

R_g —— Crystal grain
  resistance

R_b —— Grain boundary
  resistance

C_b —— Grain boundary
  capacitance

and its conduction current is related to the PN junction. The barrier height is related to temperature. When the applied voltage is greater than its reverse PN junction breakdown voltage (3.2 V), it belongs to tunnel electron breakdown conduction, and its on-state current is only related to the applied voltage. The tunnel electron breakdown time is generally less than a few hundred picoseconds.

Figure 6.11 is the equivalent circuit of a zinc oxide ceramic varistor. Among them: when the applied voltage is less than the breakdown voltage of the reverse PN junction, R_b is much greater than R_g, and the applied voltage is almost all applied to the grain boundary, at this time R_b is greater than 10MΩ. When the applied voltage is greater than the breakdown voltage of the reverse PN junction, tunneling electrons are generated in the grain boundary to conduct electricity. At this time, R_b is much smaller than R_g. Voltage is applied to the crystal grains and grain boundaries, and the unit of the sum of R_g and R_b resistance values is the ohm level. Therefore, when the applied voltage is less than the breakdown voltage (i.e., varistor voltage) of the grain boundary of the zinc oxide varistor, the varistor will exhibit a high resistance value of the insulator, and its leakage current is only microampere. When the applied voltage is greater than the breakdown voltage of the zinc oxide varistor ceramic grain boundary (that is, the varistor voltage), the varistor exhibits a low conductor resistance, and the current passing through is tens of amperes. And as the applied voltage increases slightly, the passing current increases rapidly. Figure 6.12 shows the typical V–I characteristic curve of zinc oxide varistor ceramics.

Chip zinc oxide varistor is a polycrystalline semiconductor ceramic element made of zinc oxide varistor ceramic material through electronic ceramic tape casting process. Because chip zinc oxide varistors are used in electronic circuits and data transmission lines, the working voltage of the protected circuit is very low, and there are special requirements for its capacitance. Therefore, through special design of its structure and adjustment of the process, varistors with different line protection requirements can be obtained. The internal structure and equivalent circuit of a general chip varistor are shown in Fig. 6.13.

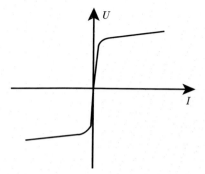

Fig. 6.12   Volt–ampere characteristic curve of varistor

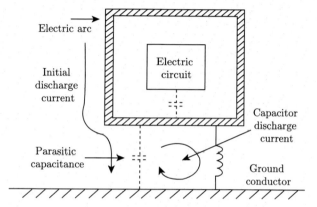

Fig. 6.13   Electrostatic discharge in a metal enclosure completely enclosed

6.2.3   ESD Suppression Method

The three most common ESD inlets are the enclosure, cable, and keyboard or control panel. The case can be metal or plastic, each with its own advantages and disadvantages. In these two cases, there are different ESD mitigation measures [16].

1. Metal Case

As shown in Fig. 6.13, the grounded metal enclosure completely seals the circuit insulated from the enclosure, and the circuit is not connected to any object outside the enclosure. Considering the large inductance of the grounding conductor, most of the initial ESD current flows through the parasitic capacitance between the enclosure and the ground. Charging leads to the rise of the potential of the enclosure, and the typical potential of the enclosure when discharging at 10,000 V is about 1000–2000 V.

An ungrounded enclosure will have a similar effect. In this case, unlike a grounded enclosure, a voltage of several thousands of volts may appear on the enclosure, which may be close to the discharge source. Therefore, in order to provide ESD protection, all metal enclosures need to be grounded.

The secondary arc can be avoided by the following ways: (1) providing sufficient space between all metal parts and circuits: (2) connecting the circuit with the metal enclosure to maintain the equipotential between the circuit and the enclosure. The space should be large enough to withstand a voltage of about 2000 V for a grounded enclosure and 15,000 V for an ungrounded enclosure.

The breakdown field strength of air at room temperature and pressure (STP) is about 3000 V/mm (75,000 V/in). The breakdown field strength is approximately proportional to the air pressure and inversely proportional to the absolute temperature. It is generally considered that the safe distance to prevent arc is one third of the above value. Therefore, for a circuit enclosed in a complete conductor enclosure and connected to the enclosure, the main electrostatic discharge problem lies in the cable interface. Therefore, measures must be taken for these cables to avoid the harm caused by electrostatic discharge.

2. Handling of Input/Output Cables

The cable becomes an ESD population in three ways: ① direct discharge; ② act as an antenna; and ③ the typical ESD problem of two boxes discussed in the previous section. The population of ESD can be prevented, or at least minimized, by one or more of the following methods.

(1) Use shielded cable;
(2) Common mode choke;
(3) Transient voltage suppression diode;
(4) Cable bypass filter.

The use of high coverage braided shielded cable or metal foil with braided layer shielding cable can protect against electrostatic discharge. The best protection is 360° welding when the shield ends on the casing. The problem of metal foil shielded cable used in ESD protection is that the shielding layer is terminated with braided wire (pigtail connecting wire), instead of 360° connection with the casing.

Figure 6.14 shows the importance of correctly connecting the shield in ESD protection. Here, consider 1 typical ESD of two boxes, which are connected together by shielded cables. From a certain point of view, this method attempts to connect two enclosures together through the shielding layer of the cable to form one enclosure. Therefore, the connection between the shielding layer and the shell is the most important parameter to determine the antistatic performance of the structure.

In addition to cable shielding, ferrite is also very effective in protecting against electrostatic discharge. The frequency range of ESD is 100–500 MHz, and it is in this frequency range that most ferrites have the highest impedance. By installing ferrite or common mode choke on the interface cable, most of the transient discharge voltage can be dropped on the choke rather than on the circuit connected to the other

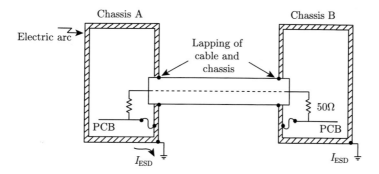

Fig. 6.14 Cable shield connecting two enclosures into a continuous enclosure

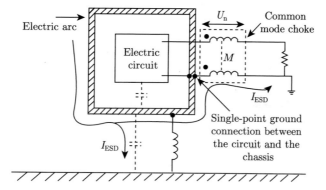

Fig. 6.15 Common mode choke installed on cable interface can reduce ESD induced noise voltage

end of the cable, as shown in Fig. 6.15. High-frequency choke or high-frequency choke discharge is necessary to minimize the high frequency of the choke. If the circuit is connected to the enclosure at only one point, the cable should enter/leave the enclosure from that point, as shown in Fig. 6.16b, and not as shown in Fig. 6.16a.

3. Insulated Enclosure

The main advantage of the insulated enclosure is that it helps to prevent the occurrence of electrostatic discharge. However, if it is not fully insulated, electrostatic discharge may also occur through gaps or holes in the enclosure.

In the case of metal enclosures, the chassis or enclosure can be used as a low-inductance path to derive the discharge current without flowing through the internal circuit. However, for non-metallic enclosure, low-inductance path does not exist, which makes ESD more difficult to control in many ways.

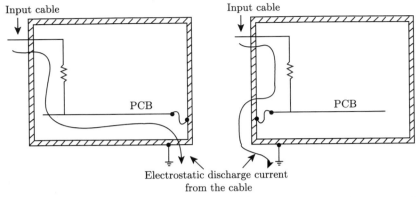

Input cable   Input cable

PCB   PCB

Electrostatic discharge current
from the cable

(a) Incorrect connection will force   (b) The correct connection can transfer the
electrostatic discharge current to flow   electrostatic discharge current from the path
through the PCB   away from the PCB to the chassis

Fig. 6.16 Connection between PCB and housing

The main disadvantages of insulated enclosures are as follows:

(1) There is no convenient alternative path for direct ESD. The green ground wire
     of AC power line does not work as the ground wire of electrostatic discharge.
(2) It can not provide shielding for indirect discharge (field coupling).
(3) There is no convenient location for connecting the following devices:

     ① Shielding layer of cable;
     ② The rear panel of the connector;
     ③ Transient voltage suppressor;
     ④ Input cable filter.

So when the product is installed in a plastic case, where should the cable shield,
transient voltage suppressor and I/O filter be connected? There are three possibilities:

(1) Connect to the ground plane of the circuit (worst choice).
(2) Connect to a separated I/O ground (better choice)
(3) It is connected to a large independent metal plate installed at the bottom of the
     product (the best choice).

When the ground of a circuit is used to transfer ESD current, a large ground voltage
will be generated, which can cause circuit damage or software errors, especially when
a complete ground plane is not used.

If all cables are connected to the same place on the PCB, a separate I/O ground
plane should be used to bypass the ESD cable current. However, in this case, the
ground wire of the I/O interface will not be connected to the enclosure because
there is no metal enclosure. The ESD current will flow through this independent
input/output ground plane and the capacitance between this plane and the actual

Fig. 6.17 Transferring ESD current from PCB using grounded metal plane

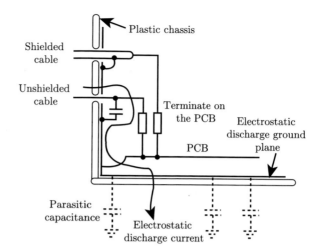

ground, not through the circuit. The effect of this method is a function of the area of input/output grounding plane and the capacitance between it and ground.

However, the best way is to install an independent ESD grounding plane in the system, which serves as a low-inductance path for both reference potential and ESD current. It is the free space capacitance of this plane that provides the ground. This line layout is shown in Fig. 6.17. When using this method, the common problem is how large the ESD grounding plane should be. The answer is simple: the ESD ground plane should be the same size as the enclosure to maximize its capacitance to ground. The electrostatic discharge plane does not need to be very thick or heavy, but it should be very large. The metal foil can play a good role, and the conductive coating on the inner surface of the plastic shell can also be used.

4. Keyboard and Control Panel

The keyboard and control panel must be designed in such a way that no discharge occurs, or even if the discharge occurs, the current can flow through the alternative path instead of directly flowing through the sensitive element. In many cases, metal spark arresters can be placed between the keyboard and the circuit to provide an alternative path for the discharge current, as shown in the figure. If it is a metal enclosure, the spark arrester should be connected to the enclosure; otherwise it should be connected to an independent ESD metal grounding plate.

Other protection methods are shown in Fig. 6.18. Figure 6.18b shows the use of an insulated shaft and/or a large knob to prevent discharge from occurring on the controller or potentiometer. Figure 6.18c shows the use of a seamless layer of insulation across the keyboard. The configuration shown in Fig. 6.18a provides an alternative path for ESD current, while the configuration in Fig. 6.18b and Fig. 6.18c prevents discharge.

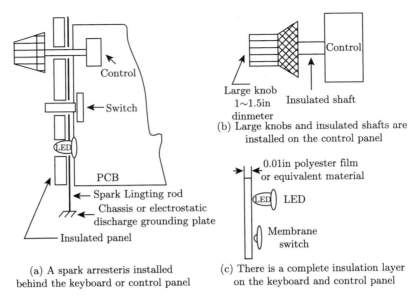

(a) A spark arresteris installed
behind the keyboard or control panel

(b) Large knobs and insulated shafts are
installed on the control panel

(c) There is a complete insulation layer
on the keyboard and control panel

Fig. 6.18   ESD suppression of keyboard and control panel

5. Hardened Sensitive Circuit

Various efforts have been made to prevent the entry of ESD current. The next step is
to reinforce the sensitive circuit. Due to the short rise time of ESD, digital circuit is
more likely to run disorderly than analog circuit. However, both analog and digital
circuits are equally vulnerable to ESD hazards.

Reset, interrupt, and any other control input that can change the operation state of
the equipment should be protected to avoid false triggering caused by rapid rise time
and short transient pulse of electrostatic discharge. These problems can be solved by
adding a small capacitor or a resistor/capacitor (or ferrite/capacitor) network (50–
100 Ω, 100–1000 pF) to reduce its sensitivity to the transient spike and narrow pulse
generated by ESD. This method should become the standard design method.

Multilayer PCB provides an order of magnitude or more ESD immunity than
double-sided PCB. These improvements are the result of using power planes and
grounding planes to reduce the impedance of power supply and grounding. Using
power planes and grounding planes also reduces the area of signal loops.

6. Electrostatic Discharge Grounding

The first thing I remember about ESD grounding is that the green ground wire of AC
power line is a high-impedance line at ESD frequency. As mentioned above, a general
conductor has an inductance of about 15 nH/in. Therefore, the grounding conductor
in a 6 T long AC power line has an inductive reactance of 20,000 at 300 MHz. This
does not include the inductance of all AC power lines in the building wall behind
the socket outlet, that is, the inductance of the conductor before the green ground
wire is actually grounded. This part of the conductor can easily reach 50 ft or longer,

which will add an additional 17,000 Ω impedance to the ground wire, making the total power line–ground impedance close to 200,000 Ω at 300 MHz.

The actual or reference ground of ESD is the ESD grounding plate in the cabinet (or metal enclosure) or product, and its free space capacitance, as shown in Fig. 6.19. If the enclosure or ESD grounding plate only has a capacitance of 26 pF, the impedance at 300 MHz is about 20 Ω. This shows that it is three orders of magnitude smaller than the impedance of the AC power line.

Therefore, the way to transfer the ESD current from the circuit is to install the transient voltage protector and/or filter on the cabinet. If there is no metal case, the transient protector or filter is mounted on a separated ESD grounding plate inside the product. If there is neither a metal case nor an ESD ground plate, all you can do is try to limit the discharge current with a resistor or ferrite and connect the protector or filter to the ground plane. If you do not have a ground plane, then you are really in trouble.

If a large amplitude of ESD current is conducted to the earth, a strong magnetic field will be generated near or inside the product, which will have adverse effects on the system. Therefore, in some cases, it is beneficial to add some resistance between ESD entry point and ground to reduce the discharge current amplitude. This is often referred to as "soft ground." In general, 100–1000 Ω is enough.

7. Ungrounded Products

Can electrostatic discharge occur on ungrounded objects? The answer is yes. We may all have seen discharge on ungrounded door handles. In this case, what is the path of ESD current? On products without external grounding connection (such as a hand-held calculator), the ESD current path will flow from the entry point to the ground through the part of the product with the maximum capacitance (minimum impedance) to ground. In many small handheld products, the part with the largest capacitance to ground is the printed circuit board. When there is ESD current flowing through PCB, the expected output will not be obtained.

The solution is to provide an alternative path for ESD current with low impedance (large capacitance) to ground. This is usually achieved by adding an ESD grounding plate under the PCB of the product. This plate blocks the capacitance between the actual PCB and the ground, and at the same time, the plate itself and the ground constitute a large capacitance for electrostatic discharge current to pass through. This is similar to the case with a metal case. This method is used on many small handheld devices such as calculators, thus providing protection against electrostatic discharge.

Figure 6.19 shows the internal structure of a plastic clamshell for a handheld calculator. The picture on the left shows the printed circuit board, and the picture on the right shows the stainless steel electrostatic discharge grounding plate installed at the bottom of the cabinet. The picture on the left of Fig. 6.19 can also see the bottom of the second metal plate set between the PCB board and the keyboard. This is a spark arrester and is used to provide an optional path for the transient current generated by discharging the keyboard.

Fig. 6.19 Flip case of handheld calculator

6.3 ESD Case Study 1

EMC Standard for Products: People's Republic of China Pharmaceutical Industry Standard (YY-0505-2012).

6.3.1 Product Introduction

The Doppler blood flow detector is a kind of desktop equipment, which mainly includes plastic case, operation panel, digital display, and probe. The detector uses a 220 V AC power supply to connect to transformer, and then to the power board after voltage reduction. The power board integrates rectifier circuit, filter circuit, and multiple relay devices to supply power for the whole system. The digital panel processes the commands from the program through the single-chip microcomputer and controls the pulse generator to convert the digital signal into the pulse analog signal. At the same time, the information is displayed on the digital display.

Figure 6.20 shows the detector.

6.3.2 Problem Description

According to the standards of the People's Republic of China (YY-0505-2012), contact discharge should pass 6 kV ESD test. The electrostatic generator of Jiangsu

Fig. 6.20   Doppler blood flow detector appearance

Electrical Equipment Electromagnetic Compatibility Engineering Key Laboratory was used to test. The Doppler blood flow detector could not pass the contact discharge test with 6 kV and the air discharge test with 8 kV. When the 6 kV contact discharge was conducted, the display indicator of the Doppler blood flow detector was off. When the 8 kV air discharges were conducted, the indicator lights of the Doppler blood flow detector display were out of order.

6.3.3   Diagnosis and Analysis of Problems

The shell of the equipment is made of PVC insulating material. The shell is complete without obvious holes and seams. There is a display on the upper surface, some metal parts on the surface, and bare metal screw on the back surface. When the shell is opened, there is a PCB with a display inside. The inner shell is made of metal and covered with insulating paint. There are many chips in the circuit without electrostatic protection measures. When conducting contact electrostatic discharge test, exposed metal makes high- frequency ESD current into PCB circuit. Without ground, it cannot make ESD current into the ground, and there is no device to block ESD current into the chip, so the chip cannot work normally, and the equipment cannot pass the test. When the air discharge test was conducted, ESD current into PCB circuit by the induction of electromagnetic field, resulting in abnormal operation of the equipment.

6.3.4   Modification Measures and Theoretical Analysis

1. Apply Antistatic Film to the Surface of the Display

Antistatic film is not electrostatic film, the electrostatic film is a kind of non-coated film, adhering to product by electrostatic. The electrostatic film is generally used for the surface sensitive to adhesive or glue residue, such as glass, lenses, high-gloss plastic surface, acrylic, and other smooth surfaces. The electrostatic film is self-adhesive and has low adhesion. The antistatic film is made of PE by adding antistatic agent to make the surface resistance reach 10–10 Ω without static electricity. After high-temperature processing, it is durable and wearable and has a good antistatic effect.

Antistatic protective film is a kind of release film. It uses an antistatic polyester film as a substrate and is coated with an acrylic adhesive and an antistatic layer. On the one hand, the antistatic agent can effectively absorb the "moisture" on the surface of the product and the air and can neutralize the static electricity when the electrostatic agent obtains moisture. On the other hand, insulating air is used as a dielectric electrolyte to improve the antistatic ability of the shell.

In electrostatic neutralization, part of the electric charge on the surface of the product is absorbed and neutralized by the antistatic agent, so that Q decreases, while C remains unchanged, according to the formula:

$$C = \frac{Q}{U} \tag{6.2}$$

By transformation of Eq. (6.2), there is [1, 2]

$$U = \frac{Q}{C} \tag{6.3}$$

When the electric charge Q on the product surface decreases, the voltage formed by the charge accumulation in the shell of the trioxide therapeutic instrument is also greatly reduced, and the voltage withstanding ability of the instrument is improved.

Because the antistatic film is added to the surface of the display, the capacitance is defined by using the air between the antistatic film and the surface of the display as a dielectric electrolyte.

$$C = \varepsilon S/d \tag{6.4}$$

where ε is the dielectric constant between the plates, S is the area of the plates, and d is the distance between the plates. When ε increases, C increases. According to the formula $U = Q/C$ and electrostatic neutralization, when Q decreases, C increases, then U decreases further and the anti-voltage ability of equipment is improved significantly. Figure 6.21 shows the measure of adding antistatic film on the surface of the display screen. From the test results with and without adding antistatic film in Table 6.4, it can be concluded that this measure is effective and the equipment has passed 8 kV air discharge test.

2. Parallel TVS Between the Display Board MCU Power Supply, Digital Tube VCC, and Ground

Paralleling TVS between VCC of the single-chip microcomputer and ground. TVS is connected in parallel between switch signal line, sensor signal line, display board

Fig. 6.21 Doppler flow detector with antistatic film on the display screen surface during ESD modification

Table 6.4 Contrast with/without antistatic film on the surface of display screen

Without adding antistatic film on the surface of the display screen	With adding antistatic film on the surface of the display screen
8000 V air discharge, indicator lamp flash disturbance	8000 V air discharge, the indicator lamp of the blood flow detector is normal

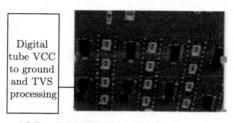

Digital tube VCC to ground and TVS processing

(a) Parallel TVS between VCC and Ground

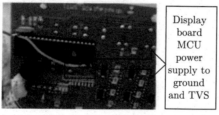

Display board MCU power supply to ground and TVS

(b) Parallel TVS between the power supply of the single chip microcomputer power and ground

Fig. 6.22 Parallel TVS between the power supply of display board, the VCC of the digital tube, and ground

MCU and ground. The clamping voltage of the TVS is 5 V. The 6 kV contact discharge test of the EUT is passed, and the green light of the display module works normally. Parallel TVS between the power supply of display board, the VCC of the digital tube, and ground and contrast without and with paralleling TVS are shown in Fig. 6.22 and Table 6.5.

Table 6.5 Contrast without and with paralleling TVS

Without TVS (6 kV)	With TVS (6 kV)
Indicator lights out	Indicator lamp working normally

Table 6.6 Comparison before and after modification

Before	After
6 kV contact discharge, indicator lights out,	Indicator lights work normally,
8 kV air discharge	indicator lights work disorderly

6.3.5   Final Modification Results

Based on the above measures, the Doppler flow detector passed the test of the 6 kV contact discharge and the 8 kV air discharge under the standard of GB/T17626. The final test results are given in Table 6.6.

6.3.6   Summary

After the measures such as coating the insulation antistatic paint on the shell, paralleling TVS between the display board single-chip power supply, the VCC of digital tube, and ground, the 6 kV contact discharge test and the 8 kV air discharge test passed the YY-0505-2012 standard.

6.4   ESD Case Study 2

Aiming at the ESD problem of the pelvic therapy device, this paper analyzes the shell and internal PCB circuit of the therapy device and proposes suppression measures for the grounding of the casing with a hole shield, the shielding of the sensitive chip, the power port of the chip, and the TVS array. The instrument has passed GB/T 17626 standard ESD test experiment, verifying the practicability of the ESD protection method.

6.4.1   Problem Description and Diagnosis

According to the pharmaceutical industry standard of the People's Republic of China (YY-0505-2012), the coupling discharge should pass the 8 kV electrostatic discharge

Fig. 6.23 Screen of the pelvic therapy device freezes during the 4 kV discharge test

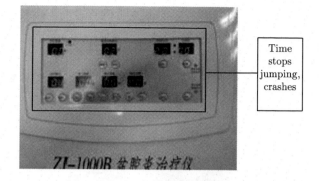

Time stops jumping, crashes

Fig. 6.24 Appearance of pelvic therapy instrument

test. The electrostatic gun of Jiangsu Xuanmiao Technology Co., Ltd. was used for ESD test. This device failed the 4 kV contact discharge on the metal part of the surface and the 4 kV air discharge on the non-metal part. The time on the display stopped beating after several electrostatic gun discharges. A crash occurs, as shown in Fig. 6.23.

The appearance of the pelvic therapy instrument is shown in Fig. 6.24. The equipment shell is made of metal material and covered with an insulating paint layer. The shell is covered with a complete shell without obvious holes. There are multiple displays on the upper surface, and some metal accessories are connected to the back. Bare metal screws, the power inlet end is provided with holes for cables to enter and exit, at this time the shielding case has some holes. After the shell is opened, there is a PCB circuit board with a display inside. The inner shell is made of metal and covered with an insulating paint layer. There are many chips in the circuit, but it lacks electrostatic protection measures. During the contact electrostatic discharge test, the ESD pulse current is easily introduced into the PCB circuit through the holes and gaps, so the chip with the ESD pulse current cannot work normally, and the device cannot pass the test. During the air gap discharge–charge coupled discharge test, the ESD pulse current surges into the PCB circuit due to the induction effect of the electromagnetic field, causing the device to malfunction.

Fig. 6.25 Ground wire of
the paint scraping metal rod
of the inner shell is
connected to the ground of
the outer shell

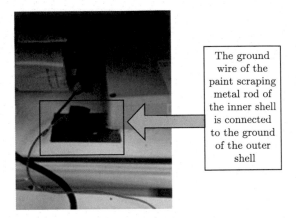

The ground
wire of the
paint scraping
metal rod of
the inner shell
is connected
to the ground
of the outer
shell

6.4.2 ESD Protection Scheme

(1) Metal Rod Grounding Treatment

Good grounding can provide a good charge discharge channel for electrostatic shock,
so that the charge accumulated on the charged body can be quickly led to the ground
avoiding sensitive devices. If the metal shield is grounded, the same amount of electric
charge induced on the outside as the charged conductor will flow into the ground,
and there will be no electric field outside the metal shell, which is equivalent to the
electric field of the charged body in the shell being shielded. According to the above
principles, the insulating paint is hung on the inner shell of the device, and then, the
inner ground wire of the metal rod is connected to the outer shell ground with copper
foil. The specific measures are shown in Fig. 6.25.

After the above-mentioned treatment, the ESD test was performed on the pelvic
therapy instrument, and the ESD immunity was significantly improved. The test
results before and after the metal rod were grounded are given in Table 6.7.

(2) Shield the Sensitive Area on the PCB with Copper Film

When the electromagnetic wave is shot on the surface of a piece of metal, its energy
intensity will attenuate to zero due to loss. Once entering the surface of the conductor,
a high-frequency AC electromagnetic field will be generated in the conductor. The
penetration depth of electromagnetic waves into a metal body is mainly affected by
its frequency and the conductivity and permeability of the conductor. The lower the
frequency, the smaller the conductivity, and the smaller the magnetic permeability,
the deeper the penetration depth, and vice versa. According to the above principle,
part of the sensitive area on the PCB where the display is located is covered with
copper foil and shielded, as shown in Fig. 6.26.

After testing, when the equipment is shielded as above, its air discharge capacity
increased to 7 kV. Before and after the copper foil shielding, before and after the
ESD test, the results are given in Table 6.8, so the above measures are effective.

Table 6.7 Comparison of test results before and after the metal rod is grounded

Before the metal pole grounding treatment	After the metal rod is grounded
4 kV contact and non-contact discharge crash	Contact discharge can be increased to 5 kV, air discharge can be increased to 6 kV

Fig. 6.26 Attach copper film to the time chip on the PCB

Table 6.8 Test comparison before and after the copper film treatment of the time chip on the PCB board

Before copper foil treatment	After copper foil treatment
Contact discharge resistance 5 kV, air discharge resistance 6 kV	Contact discharge can be increased to 5.5 kV, air discharge can be increased to 7 kV

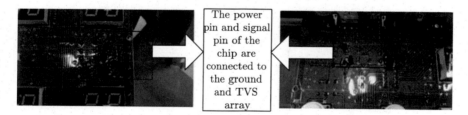

Fig. 6.27 Power pin and signal pin of the chip are connected to the ground and TVS array

(3) In the Chip's Power Pin and Signal Pin and TVS Array

The clamping characteristics of TVS can effectively keep the voltage at both ends of the chip within the safe working voltage range, but only a single TVS tube cannot achieve the ideal suppression effect, so the time signal chip on the display PCB board is used. The power pins and signal pins are connected to the ground and TVS array measures, as shown in Fig. 6.27. When the transient overshoot voltage caused by ESD interference is encountered, the large voltage coupled to the sensitive line will break down the TVS and cause large current. It has been discharged through the TVS tube before flowing through the chip to suppress electrostatic interference.

Table 6.9 Before and after comparison of the chip power pin and signal pin to ground and TVS array

Before TVS array processing	After TVS array processing
The resistance to contact discharge increases to 5.5 kV, and the resistance to air discharge increases to 7 kV	The resistance to contact discharge increases to 6 kV, and the resistance to air discharge increases to 8 kV

Table 6.10 Before and after the experiment

Before the experiment	After the experiment
4 kV contact and non-contact discharge crash	Contact discharge can increases to 6 kV, air discharge can increases to 8 kV

After testing, when the device performs the chip's power pin and signal pin to the ground and TVS array processing, its contact discharge capacity increases to 6 kV, and the air discharge capacity increases to 8 kV. Before and after the treatment, the comparison between before and after the ESD test is given in Table 6.9. As shown, the above measures are effective.

6.4.3 Experimental Results and Analysis

The pelvic therapy instrument is grounded by the metal rod, the electromagnetic field shielding of the sensitive chip of the internal PCB board, the chip power pins and signal pins are grounded, and the TVS array suppresses three measures, which significantly improves the ESD immunity of the pelvic therapy instrument. At the same time, under the GB/T17626 standard, the electrostatic gun of Jiangsu Xuanmiao Technology Co., Ltd. was used for coupling discharge test. The pelvic therapy instrument passed the contact discharge test with a voltage of 6 kV, and the air discharge with a voltage of 8 kV, and the screen showed normal. The final test result comparison is given in Table 6.10.

References

1. Nave MJ. Power line filter design for switched-mode power supplies. 2nd ed. Hoboken: Wiley; 1991.
2. Caccavo G. ESD field penetration into a populated metallic enclosure. IEEE Trans Electromagn Compat. 2002;44(1):243–9.
3. Guo T. Separation of the common-mode and differential-mode conducted EMI noise. IEEE Trans Power Electron. 1996;11(3):480–8.
4. Paul CR. Diagnosis and reduction of conducted noise emissions. IEEE Trans Electromagn Compat. 1988;30(4):553–60.

5. See KY. Network for conducted EMI diagnosis. IEEE Electron Lett. 1999;35(17):1446.
6. Zhao Y. Fundamental of electromagnetic compatibility and application. Beijing: China Machine Press; 2007. p. 174.
7. Li H, Miao M, Zhou Y, et al. Modeling and simulation of comprehensive diode behavior under electrostatic discharge stresses. IEEE Trans Device Mater Reliab. 2019;19(1):90–6.
8. Bhattacharya P, Sinha R, Thakur BK, et al. Adaptive dielectric thin film transistors—a self-configuring device for low power electrostatic discharge protection. IEEE Electron Device Lett. 2020;41(1):66–9.
9. Cai X, Yan B, Huo X. An area-efficient clamp based on transmission gate feedback technology for power rail electrostatic discharge protection. IEEE Electron Device Lett. 2015;36(7):639–41.
10. Zhang C, Liu S, Xu K, et al. A novel high latch-up immunity electrostatic discharge protection device for power rail in high-voltage ICs. IEEE Trans Device Mater Reliab. 2016;16(2):266–8.
11. Zhang R, Zhao W, Hu J, et al. Electrothermal characterization of multilevel Cu-graphene heterogeneous interconnects in the presence of an electrostatic discharge (ESD). IEEE Trans Nanotechnol. 2015;14(2):205–9.
12. Xi Y, Salcedo JA, Dong A, et al. Robust protection device for electrostatic discharge/electromagnetic interference in industrial interface applications. IEEE Trans Device Mater Reliab. 2016;16(2):263–5.
13. Liu D, Nandly A, Zhou F, et al. full-wave simulation of an electrostatic discharge generator discharging in air-discharge mode into a product. IEEE Trans Electromagn Compat. 2011;53(1):28–37.
14. Horng JJ, Su YK, Chang SJ, et al. Nitride-based Schottky barrier sensor module with high electrostatic discharge reliability. IEEE Photonics Technol Lett. 2007;19(10):717–9.
15. Lee TS, Gasal J. Partial electrostatic discharge induced by precharged dielectric surfaces. IEEE Trans Ind Appl. 1997;33(3):692–6.
16. Bogorad AL, Likar JJ, Voorhees CR, et al. Electrostatic discharge induced momentum impulse from charged spacecraft surfaces. IEEE Trans Nucl Sci. 2006;53(6):3607–9.

Chapter 7
Principle and Analysis of EMS: EFT Mechanism and Protection

7.1 Formation Mechanism of EFT

Long-term experience in electrical equipment susceptibility testing has shown that it is necessary to simulate fast transient tests with high repetition rates to investigate the susceptibility of sensitive equipment. To ensure the accuracy and comparability of test results, the IEC has developed Related Electrical Fast Transient Burst Susceptibility Test Standard IEC 6100-4-4 "Electromagnetic Compatibility Test and Measurement Technology Electrical Fast Transient Burst (EFT) Susceptibility Test" (China converts this standard equivalent into national standard GB/T 17626.4). This standard specifies the definition, working principle, measurement method, and test generator of EFT, which becomes the basis of other EFT reference [1–5]. EFT susceptibility is an important item in product EMC and has been included in most product testing standards. This section explains the formation mechanism of electrical fast transient pulses, implementation standards and test methods, and adjustment methods.

7.1.1 Mechanism and Analysis of EFT

The full name of EFT is electrical fast transient, which represents a series of transient pulse (pulse group) interference generated at the switch due to insulation breakdown of switch contact gap or contact bounce when the inductive load is disconnected and connected. Electrical fast transient is generated by inductive devices such as relays, motors, and transformers. These components form part of the system, so interference is often generated inside the system.

When the inductive load is repeatedly turned on and off multiple times, the pulse group occurs repeatedly in the corresponding time gap. The generation of such pulses includes pulses generated by switching of small inductive loads, pulses generated by relay electric shock, and pulses generated by switching of high-voltage switching devices.

© Science Press 2021
Y. Zhao et al., *Electromagnetic Compatibility*,
https://doi.org/10.1007/978-981-16-6452-6_7

Fig. 7.1 Circuit formed by
electric fast pulse

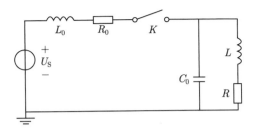

In Fig. 7.1, L_0, R_0, and C_0 represent the stray inductance, wire resistance, and stray capacitance of the circuit. In the process of the switch breaking the inductive load circuit, since the inductor current cannot be abruptly changed, this current flows to the stray capacitance C_0, which is reversely charged to form a back electromotive force.

For the transient discharge loop, the second-order dynamic circuit formula can be written according to the initial voltage and Kirchhoff's voltage law [6–8].

$$\begin{cases} LC_0 \dfrac{\mathrm{d}^2 U_c}{\mathrm{d}t^2} + RC_0 \dfrac{\mathrm{d}U_c}{\mathrm{d}t} + U_c = 0 \\ U_c(0_+) = U_c(0_+) \\ \dfrac{\mathrm{d}U_c}{\mathrm{d}t}(0_+) = \dfrac{\mathrm{d}U_c}{\mathrm{d}t}(0_-) = I_S \end{cases} \tag{7.1}$$

where $I_S = U_S/R$ is the current in the loop when the switch is closed.

Solving the circuit formula shown in Eq. (7.1), the transient voltage across capacitor C_0 is [9]:

$$U_c = \frac{U_S}{\omega C_0 R} e^{-\sigma t} \sin(\omega t) = \frac{I_S}{\omega C_0} e^{-\sigma t} \sin(\omega t) \tag{7.2}$$

where $\omega = \sqrt{\dfrac{1}{LC_0} - \dfrac{R^2}{4L^2}}$ is the free resonant angular frequency, $\tau = R/2L$ is the attenuation coefficient.

Transient overvoltage occurs at both ends of the inductive load. This overvoltage is superimposed on the power supply voltage and applied to both ends of the switch contacts. When the voltage across the contact is higher than the dielectric breakdown voltage, an arc is formed between the contacts, and the switch is reignited. After the switch is turned on, C_0 is discharged to form a high frequency current, and the reignition arc between the contacts is extinguished, and an overvoltage occurs at both ends. The above process occurs repeatedly until the voltage on the capacitor does not cause the dynamic and static contacts of the switch to break down. During the reignition and extinction of the arc, the pulse voltage amplitude is low, the rise time is fast, and the frequency is high. As the distance between the switching dynamic and static contacts increases, the amplitude of the pulse voltage gradually increases, the pulse front becomes slower, and the pulse repetition frequency decreases.

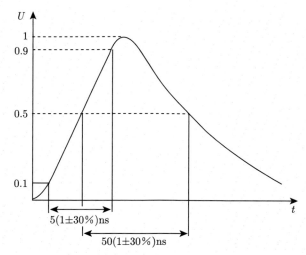

(a) Typical waveforms specified by the standard

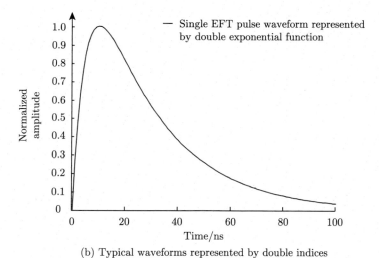

(b) Typical waveforms represented by double indices

Fig. 7.2 Single EFT pulse waveform

Figure 7.2a shows the typical waveform of the unit electrical fast transient. Combined with the description of Eq. (7.3), a single electrical fast transient pulse can be defined as a double exponential pulse signal source, and Eq. (7.3) is a double exponential single pulse expression [10].

$$U(t) = k_p U \left(e^{-t/\tau_1} - e^{-t/\tau_2} \right) \tag{7.3}$$

where U is the voltage amplitude in volts, k_p is the compensation coefficient of the standard EFT pulse, τ_1 is the wave front coefficient in ns, τ_2 is the wavelength coefficient in ns.

In order to satisfy the rise time of a single pulse $t_r = 5\,(1 \pm 30\%)$ ns, the standard EFT pulse with a hold time of 50 $(1 \pm 30\%)$ ns, the values of Eq. (7.3) are: $k_p = 1.875$, $\tau_1 = 25$ ns, $\tau_2 = 5$ ns, U is normalized to indicate that a single pulse obtained by taking 4 kV/2 kV/1 kV/0.5 kV according to the standard is shown in Fig. 7.2b. It is basically consistent with Fig. 7.2a and meets the requirements of the standard waveform. At the same time, the duration of a single pulse is much smaller than the duration of the entire burst, so the interaction between the pulses can generally be ignored.

7.1.2   EFT Interference Mechanism

When the switching system of inductive load is disconnected, transient interference can occur at the disconnection point, which is composed of many pulses. For the 110 V/220 V power supply line, the measurement shows that the range of the EFT amplitude from 100 V to thousands of volts, the specific value is determined by the electromechanical characteristics of the switch contacts, and the range of the pulse repetition frequency from 1 kHz to 1 MHz. For a single pulse, the rising edge is in the nanosecond order, and the duration of the pulse ranges from tens of nanoseconds to milliseconds. The spectrum of the interference signal is widely distributed. Therefore, digital circuits are sensitive to it and are susceptible to harassment.

Based on the standard of GB/T17626.4—2018, the EFT is characterized by a pulse group rather than a single pulse. A single pulse has a short rise time and low energy, and it does not damage the EUT. However, repetitive high-frequency pulses group can cause disoperation of devices or control chips due to the constant charging of junction capacitors of EUT semiconductor devices, which leads to the failure of electronic products.

The simplified structure diagram of the EUT of single-phase three-wire system is shown in Fig. 7.3. On the left side of Fig. 7.3 is an EFT generator, which can generate pulses in the range from 5 kHz to 100 kHz specified in the standard GB/T17626.4—2018. There are three 50 cm power cables to transmit the pulse group to the power port of EUT. At this time, there are three kinds of interference in the EFT test for the EUT.

The first one is coupling through conduction on the transmission line. When EFT interference is imposed on the power line through the coupling network, one end of the output of the EFT is injected into the measured power line through the capacitance of 33 nF, and the other end is connected to the ground through the ground terminal of the coupling unit. When it imposes interference on the signal/control line, the EFT enters the EUT through the interference capacitance between the coupling clamp and the cable under test. The EFT conducted in this way is a common mode noise.

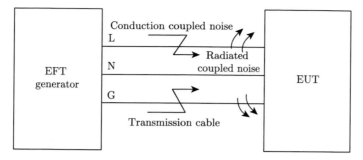

Fig. 7.3 Simplified EUT of single-phase three-wire system

The second one is near-field radiated noise caused by transmission line. The rising edge t_r of the single pulse of EFT reaches 5 ns, which results in many harmonic components of it. These harmonic components radiate into space when passing through 50 cm transmission cables. This radiation energy is coupled to adjacent cables, and through these cables, near-field EMI interference is generated in the circuit. This space radiation also enters the power line of the EUT in the way of a typical common mode.

The third one is the superposition of the above two kinds of interference. During the transmission of signals on cables, some noise interference from the conducted interference noise coupled on cable, while the other part escapes from the cable and becomes the radiated noise entering the EUT, and causing near-field radiation interference.

Because of the above three interference ways, the noise source of the EFT becomes complex, which directly affects the difficulty and cost considerations of the design, manufacture, certification and other stages of electrical and electronic products. Among them, the transmission of noise is also related to the coupling of electromagnetic field in space, which makes the mechanism analysis more difficult.

7.2 EFT Protection Theory

7.2.1 EFT Standardization and Testing Methods

IEC 61000-4-4 is a standard specially formulated for EFT. Because of the wide range of applications, many international organizations and some relevant departments in China have introduced this standard as the testing standard for their products. The definition of EFT is shown in Fig. 7.4. According to the Electrical Fast Transient Pulse Group Susceptibility Test, EFT is a series of transient impulse (pulse group) interference generated by the insulation breakdown or contact bounce of the switch contact gap when the inductive loads in the switching process. When the induc-

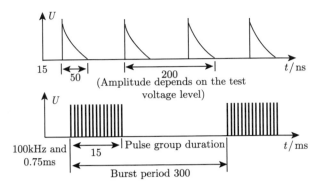

Fig. 7.4 Waveform of EFT

tive load repeats the switching process, the EFT repeats in the corresponding time gap. These pulses include the pulses generated by the switching of small inductive loads, the pulses generated by the shock of electric relay (which mainly interfere with sensitive devices by conduction), and the pulses generated by the switching of high-voltage switchgear (which mainly interfere with sensitive devices by radiation). The remarkable characteristics of this kind of transient interference are fast-rising time, short duration, low energy but high repetition frequency. This kind of transient interference has low energy and does not cause damage to the equipment in general. However, because of fast-rising time and repetition frequency, the spectrum distribution is wide, so there is an impact on the reliability of electronic and electrical equipment.

The parameters of the standard are usually typical values, which conform to the statistical law. The repetition frequency of the EFT in the actual electromagnetic environment ranges from 10 kHz to 1 MHz. If the rise time is measured near the source of the EFT, it is almost the same as the pulse generated by electrostatic discharge through air (about 1 ns). If the measurement site is far away from the source of the EFT, the rise time is prolonged due to transmission loss and reflection. This factor is taken into account when the length of cable between the EUT and the source of the EFT is less than 1 m. The rise time of the standard is 5 ns, which is the comprehensive value after considering many factors.

1. Test Equipment

The test equipment is as follows: ① Grounding reference plane: The grounding reference plane should be a metal plate (copper or aluminum) with a minimum thickness of 0.25 mm. Other metal materials can also be used, but their minimum thickness should be 0.65 mm. The minimum size of the ground plane is 1 m × 1 m, and the actual size is related to the size of the EUT. ② Coupling device (network or capacitance coupling clamp). ③ Test generator. The photograph of the EFT interference test is shown in Fig. 7.5.

Fig. 7.5 Photograph of the EFT interference test

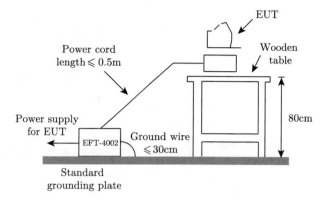

Fig. 7.6 EFT configuration test for power line of desktop equipment

2. Test Configuration

The ground plane shall extend at least 0.1 m beyond the area of the EUT and be connected to the protective ground. Except for the ground plane below the EUT, the minimum distance between the EUT and all other conductive structures (such as the walls of the shielded room) is greater than 0.5 m. The test is provided with a grounding cable that is connected to the ground reference plane and all connectors to minimize the inductance provided. The length of the signal and power lines between the coupling device and the EUT should be 1 m or less. If the length of the power lines of the equipment exceeds 1 m, the excess shall be gathered together to form a flat coil of 0.4 m in diameter and located 0.1 m above the ground reference plane. The distance between the EUT and the coupling device shall be kept at 1 m or less.

3. Several Implementation Methods of Typical EFT Testing

1. EFT test of Desktop Equipment

    (1) The EFT configuration test of power line (Fig. 7.6).
        The EUT is connected to the grounding system according to the installation requirements of the product manual and does not allow additional grounding. If the length of the power line exceeds 1 m, the power line should be bent

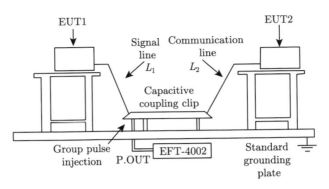

Fig. 7.7 EFT configuration test for signal line of desktop equipment

into a flat loop with a diameter of 40 cm, and then placed parallel to the reference grounding plate with a height of 10 cm.

(2) The EFT configuration test of signal line (Fig. 7.7).

A capacitive coupling clip is used for the test. If the test is conducted on two EUT, the length of signal lines L_1 and L_2 exposed between the equipment and the capacitive coupling clamp must be less than 1 m. If the test is conducted on a EUT (e.g., EUT 1), the L_1 must be less than 1 m, while the L_2 should be more than 5 m, and the extension part of L_2 should be bent into a flat loop and placed parallel to the reference grounding plate with a height of 10 cm.

2. EFT Test of Vertical Equipment

(1) The EFT configuration test of power line (Fig. 7.8).
The requirements of connection are the same as desktop equipment.
(2) The EFT configuration test of signal line (Fig. 7.9).
The requirements of connection are the same as desktop equipment.

3. EFT Test for Power Supply of Desktop Equipment in the Field
When the EFT Test for power supply of equipment or system is conducted in the field, in order to simulate the field interference as realistically as possible, the coupling/decoupling network should not be used in the test and the EFT is directly performed through the 33 nF coupling capacitor. (Apply a pulse directly to the power line of the EUT and the length of the power cable is 1 m.) If other equipment or systems are also affected during the test, a coupling/decoupling network such as EFT-4002 can be considered to ensure the reliability of other equipment. The EFT test for power supply of benchtop non-stationary equipment is shown in Fig. 7.10. The test voltage should be applied between each power line and the protective ground line connected to the EUT.

4. EFT Test for Power Supply of Vertical Equipment in the Field
The EFT test for power supply of vertical equipment in the field is shown in Fig. 7.11. Place a 1 m × 1 m reference grounding plate near the EUT. The

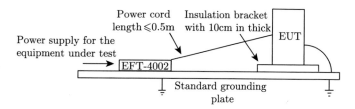

Fig. 7.8 Configuration of EFT test for power line of vertical equipment

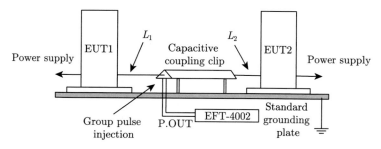

Fig. 7.9 Configuration of EFT test for signal line of vertical equipment

EFT generator is placed on the ground plane, and the shell is connected to the ground plane via a short, thick ground line. The ground plane is connected to the protective ground of the main power supply.

The EFT test principle for power supply of vertical equipment is the same as the test of desktop equipment in the field. During the test, the EFT is output from the coaxial terminal POUT to the test point of the EUT. The connection is not necessarily shielded, but good insulation is required. The length of it is not more than 1 m. If an AC/DC isolation capacitor is to be used, the capacitance should be 33 nF. Other power connections of the EUT should meet the functional requirements.

5. EFT Test of I/O and Telecommunication Line in the Field

The EFT test of I/O and telecommunication line in the field is shown in Fig. 7.12. When conducting EFT test of I/O lines and telecommunication lines in the field, capacitive coupling clamps should be employed to couple interference signals to the lines. If the coupling clamp cannot be employed because of the size and laying of the signal or telecommunication line, it can also be wrapped with metal foil or strip to replace the coupling clamp, but the coupling capacitance produced should be equivalent to that of the capacitor clamp. At this time, the coaxial output core POUT of EFT-4002 is connected with metal foil, and the shielding layer of the cable is connected with the shell of the EUT. It is worth noting that the connection should be as short as possible.

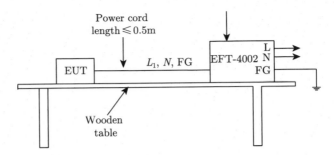

Fig. 7.10 EFT test for power supply of benchtop non-stationary equipment

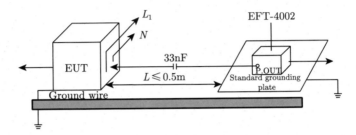

Fig. 7.11 EFT test for power supply of vertical equipment in the field

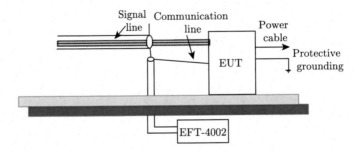

Fig. 7.12 EFT test of I/O and telecommunication line in the field

7.2.2   EFT Restraint Measures

1. Reduce the Common Resistance on the PCB Ground Line

The EFT interference signal is usually transmitted into the interfered device from the power line or signal line, and it is a group of rapidly changing pulse signals. If the interfered device does not have good filtering performance at the power source or signal input and output ends, part of the EFT signal will enter the subsequent circuit of the interfered device. Few modern electronic equipment does not contain digital circuits, and digital circuits are more sensitive to pulse interference. The EFT signal that intrudes into the subsequent circuit will cause the digital circuit to work abnormally through direct triggering or spatial coupling. Figure 7.13 shows the

Fig. 7.13 EFT charges IC parasitic capacitance

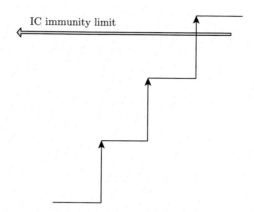

IC immunity limit

situation where the digital circuit is interfered by spatial coupling. At the input end of the IC, the EFT charges the parasitic capacitance, through the gradual accumulation of numerous pulses, and finally reaches and exceeds the IC's immunity limit.

In addition, the intrusive EFT signal will also interfere with the device under test through the common ground wire of the PCB. The ground wire here refers to the connection wire of each circuit and unit potential reference in the electronic device, that is, the signal ground wire. Since any ground wire has both resistance and reactance, a voltage drop will inevitably occur when current flows. For EFT signals, the current changes extremely fast and contains a lot of high-frequency components. According to:

$$U = -L\frac{\mathrm{d}i}{\mathrm{d}t} \tag{7.4}$$

It can be seen that it is easy to produce a potential difference on the common ground, and the voltage is proportional to the L value and the $\mathrm{d}i/\mathrm{d}t$ value. If the voltage drops below the noise immunity level of the digital circuit, there will be no interference problems; otherwise, it may cause interference to other circuit units sharing the ground wire.

2. EFT Inductive Transient Interference Suppression Network

Figure 7.14 shows the six network structures that suppress EFT interference in a circuit with inductive load.

(1) As shown in Fig. 7.14a, a resistor is connected across the inductor. When the switch is turned off, the inductor causes the current that originally passed through it to flow through the resistor. The transient voltage peak value increases with the increase of resistance, but it will not exceed the steady-state current multiplied by the resistance. If $R = R_L$, the transient voltage is equal to the power supply voltage, and the voltage on the contact is equal to the power supply voltage plus the induced voltage on the coil. The efficiency of this circuit is very low, because the resistance consumes energy. If $R = R_L$, the resistance and the load consume the same energy.

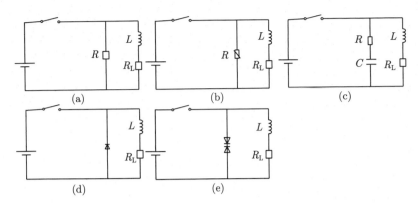

Fig. 7.14 Inductive EFT interference suppression network

(2) As shown in Fig. 7.14b, a varistor is used instead of an ordinary resistor. When the voltage on the varistor is lower, its resistance is larger, but when its voltage is higher, its resistance is smaller. The cutoff voltage of the varistor is higher than the working voltage.

(3) As shown in Fig. 7.14c, a resistor–capacitor series network is connected across the inductor. When the circuit works in a steady state, no energy is consumed. When the contact is disconnected, the capacitor will initially be a short circuit, and the inductor will drive current flow through the resistance. When the contact is closed, the inductance impedance is very large, and the current mainly flows through the resistance. At this time, the resistance must limit the current. The larger the resistance, the better the current limiting effect when the switch is closed, so that arc discharge cannot occur. This requires a larger resistance. When the contact is disconnected, the resistance must provide an energy discharge path. The smaller the reverse charging voltage when the switch is disconnected, the smaller the better. The compromise is: $U_{peak}/I_A < R < R_L$, R_L is the arc current.

(4) As shown in Fig. 7.14d, a semiconductor diode is connected across the inductor. The diode is connected so that no current flows when the short circuit is in steady state, but when the switch is off, the voltage on the diode is opposite polarity and the diode begins to conduct, so that the voltage on the inductor is limited to a small value. The disadvantage of this circuit is that the current in the inductor decays very slowly. If the inductor is a relay, the release time will be extended. The diode peak direction voltage is higher than the power supply voltage, and the forward current is greater than the load current. If a resistor is connected in series with the diode, the release time of the relay can be shortened, but it also limits the role of the diode.

(5) As shown in Fig. 7.14e, using a Zener diode in series with the rectifier diode can shorten the decay time of the current on the inductor, because when the voltage is lower than the turn-on voltage of the Zener diode, the current is interrupted. The voltage on the contacts is equal to the supply voltage plus the voltage on the Zener diode.

(6) As shown in Fig. 7.14, it can be used for communication occasions.

3. Other Ways to Suppress EFT

Based on the characteristics of EFT interference, some targeted countermeasures are proposed above. These countermeasures are mainly to take corresponding suppression measures on the EFT injection port, shell, and interface. In addition to these external suppression measures, it is also very necessary to improve the anti-interference ability of the internal circuit of the device under test. How to improve the internal anti-interference ability of electronic products is a very important content in product electromagnetic compatibility design. This is the main part of all electromagnetic compatibility design books and articles. I will not introduce it in detail here, and mainly focus on the characteristics of EFT interference suppression in the internal circuit of the device. The following design points are proposed.

(1) It is recommended to digitize the analog signal transmission port from the PCB, adopt balanced transmission or use transformer isolation;

(2) It is recommended to use optocoupler isolation or transformer isolation for digital signal lines from PCB, or directly change optical fiber transmission;

(3) In analog circuits, symmetrical balanced amplifiers have stronger common mode interference suppression capabilities than unipolar amplifiers;

(4) In digital circuits, all unused input ports are connected to the ground or power supply and cannot be left floating;

(5) For smart chips, level trigger is much stronger than edge trigger to resist pulse interference;

(6) For the interface connected with the outside, the interface chip with gating function has stronger anti-interference ability than the interface chip without gating function;

(7) For intelligent circuits with CPU, it is necessary to add anti-jamming instructions in the software and adopt "watchdog circuit";

(8) Do not allow external signal lines to enter/exit the CPU directly without isolation by the interface chip.

7.3 Principle and Analysis of EMS: EFT Case Analysis

EMC Standard for Products: People's Republic of China Standards (YY-0505—2012).

7.3.1 Analysis of EFT Case 1

1. Product Introduction and the Description of Problem

The picture of the oxygen machine is shown in Fig. 7.15. It is a kind of machine which produces oxygen.

Fig. 7.15 Outward appearance of oxygen machine

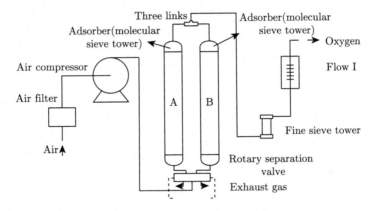

Fig. 7.16 Structure of the oxygen machine

The structure of the oxygen machine is shown in Fig. 7.16.

Standard GB/T 17626.4—2018 Electromagnetic Compatibility Test and Testing Technology Electrical Fast Pulse Group Anti-Jamming Test (equivalent to Standard IEC 61000-4-4 standard) specifies the test standard and test method of EFT as shown in Table 7.1.

Owing to the application of radio frequency devices, switching devices and digital circuit control module in the system, oxygen machine not only produces a lot of electromagnetic interference but also be affected by external electromagnetic environment noise. According to the standard of the People's Republic of China (YY-0505—2012), the EFT generator is used at the power supply port and conduct the EFT, the peak voltage of the EFT is positive or negative 1 and 2 kV, the repetition rate is 5 kHz, and the test time is 60s. In normal operation, after the operation signal is given to the oxygen machine from the operation panel, the thin white smoke can be generated from the oxygen tank, and the program response is timely and sensitive. The oxygen machine failed to pass the EFT test after conducting the EFT to the power

Table 7.1 Test rank of EFT

Rank	Power port and protective grounding (G)		I/O signal, data and control ports	
	Peak voltage (kV)	Repetition frequency (kHz)	Peak voltage (kV)	Repetition frequency (kHz)
1	0.5	5/100	0.25	5/100
2	1	5/100	0.5	5/100
3	2	5/100	1	5/100
4	4	5/100	2	5/100
X	Given	Given	Given	Given

Note X is an open level, which must be specified in special equipment technology

Fig. 7.17 Failure diagram of the oxygen machine

supply of the equipment. The specific phenomenon is that the buttons are insensitive, the display screen is splashed, and the concentration is lowered during the test. And only when the power is cut off and restarted can it be restored to normal, and it cannot be restored to normal by itself. The failure diagram of the oxygen machine when it is disturbed by EFT is shown in Fig. 7.17.

2. EFT Inhibition Theory Analysis

EFT mainly interferes with the ports of the power supply and transformer of the oxygen machine by conduction. The PCB contains many sensitive electronic components such as MOSFET and diode. When these components are impacted by the EFT, the electric charge accumulates in the nonlinear energy storage elements, which affects the function of the sensitive elements and thus interferes with the internal circuit of the oxygen machine, seriously affecting the normal operation of the oxygen machine. According to the principle analysis, first check the ground line. If the display board has a display problem when the grounding is normal, it is necessary to check the power supply line. After opening the shell, it can be seen that there are some defects in the system design. The close distance between the high-frequency switching device and the sensitive amplifier causes the amplifier to receive too much high-frequency crosstalk. At the same time, the wiring between the circuit boards is cluttered and too long, which causes the circuit to be equivalent to a radiation antenna.

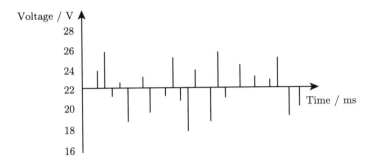

Fig. 7.18 Output waveform of oxygen machine

There is no effective shielding measure, just the simple grounding of power lines on PCB plates. There is an internal motor for generating trioxide, which is a nonlinear device that couples many interference signals of different frequency bands. These signals are not shielded, so they are received by the display module of the oxygen machine.

3. Analysis of EFT Rectification Methods

When EFT is conducted to oxygen machine, the influence of EFT on oxygen machine can be observed by oscilloscope and voltage probe. Conducting 2 kV EFT interference to the power supply port in the experiment leads to the phenomenon of peak interference voltage. The output waveform of oxygen machine after conducted EFT is shown in Fig. 7.18.

  1. Installing EFT Low-pass Filter on Power Line

The low-pass filter consists of a common mode choke, capacitor, and inductor. The mechanism of the common mode choke is that when the normal current passes through, the current forms a reverse magnetic field in the coil and cancels each other, and the normal current is mainly affected by the resistance of the coil; when the high-frequency common mode noise current passes through the coil, it generates a magnetic field in the coil in the same direction and increases the inductance of the coil, so that the coil presents a high-impedance state. Thus, the common mode choke can attenuate the high-frequency common mode current and gets the purpose of filtering. Capacitance has the function of absorbing harmonics. It is used as an electrical bypass in a low-pass filter. It matches with inductance to achieve impedance matching, which makes the filtering effect better.

Firstly, a conventional EFT filter is designed as shown in Fig. 7.19. The inductance value of mode choke coil N_1 and N_2 is 2 mH, while the capacitance value of C_1 and C_2 between neutral line and fire line is 20 nF, and the capacitance value of C_3 and C_4 between neutral line and ground line is about 5 nF.

Because of the large amplitude and wide frequency band of EFT, and the energy of EFT is mainly concentrated in 1–100 MHz, which belongs to high-frequency electromagnetic interference. The high-frequency noise of circuit can be filtered by

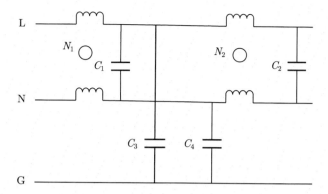

Fig. 7.19 Layout of EFT filter

Fig. 7.20 Picture of filter installation

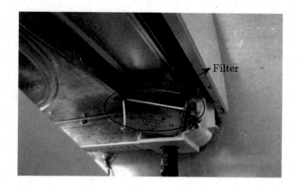

a low-pass filter. However, because of the high energy of EFT, the rising edge t_r of the single pulse waveform of the EFT is 5 ns, the pulse width is 50 ns, and the pulse amplitude is quite large, so the EFT has a very rich harmonic component. By the formulas (7.5) and (7.6), the harmonic center frequency and wavelength can be known. The conventional EFT low-pass filter cannot filter out such wideband noise, and the EUT also has screen flicker phenomenon in 2 kV EFT test. Therefore, it is necessary to improve the performance of low-pass filter. The picture of filter installation is shown in Fig. 7.20.

$$f = \frac{1}{\pi t_r} = 64\text{MHz} \tag{7.5}$$

$$\lambda = \frac{U}{f} = \frac{3 \times 10^8}{f} \tag{7.6}$$

By suppressing parasitic parameters of common mode choke, which includes self-inductance parasitic parameters and mutual inductance parasitic parameters, the effect of ground capacitance in choke at high frequency is improved, and the high-

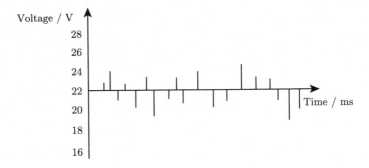

Fig. 7.21 Measurement results of EFT test

Table 7.2 Comparison of EFT test results

Measures	Test results	
	Voltage	Results
Without filter	2000 V	Not pass
Conventional low-pass filter		The effect is not ideal
Improved low-pass filter		The effect is stable and obviously improved

frequency equivalent series inductance is eliminated. The oxygen machine can work properly when the voltage amplitude of EFT is raised to 2 kV again. This result shows that the improved EFT filter works better than before. The measurement results are shown in Fig. 7.21 and satisfied. Table 7.2 shows the measurement results of EFT test.

2. Installing a Ferrite Ring on the Power Line

The ferrite ring exhibits a high-resistance state in medium- and high-frequency bands, and installing it on a power line is equivalent to series differential mode inductance on the power line. When the EUT is working, the ferrite ring makes a pulsed magnetic field generated by the high-frequency pulse current of the power line. These ferrite fields form a vortex on the ferrite ring, and the high-frequency energy is consumed in the form of thermal energy. Therefore, ferrite ring has a good inhibitory effect on the high-frequency differential mode interference signal. Different ferrite rings have different impedance characteristics, so it is necessary to calculate the cutoff frequency of the ferrite ring when designing the filter and make the cutoff frequency near the frequency of the interference signal. Ferrite ring is an absorptive filter device, multi-loop winding can change the impedance characteristics and improve the filtering performance in the whole frequency band.

Ferrite rings play a very important role in dealing with EMC problems. It cannot only filter out the interference of external environment to the system, but also prevent the interference from internal systems. It has a good inhibitory effect on

Fig. 7.22   Installation of ferrite rings

Table 7.3   Comparison of EFT test results

Measures	Test results	
	Voltage	Result
Without ferrite ring	2000 V	Not pass
With single-turn coil magnetic ring		There is a slight jitter in the display
With multi-turn wire-wound ferrite ring		The jitter in the display is improved significantly

Table 7.4   Contrast table before and after EFT amendment

Voltage grade	Before	After
1 kV	Reduced button sensitivity and abnormal flashing of indicator	The button responds normally, the indicator lamp glows normally, and there is no flicker
2 kV	Abnormal blurred image with buzzing on screen	There were no blurred images on the screen and no buzzing

high-frequency interference. The installation of ferrite rings is shown in Fig. 7.22. Table 7.3 shows the comparison of EFT test results.

4. Final Results After Modification

After the above analysis and modification, the ferrite ring and EFT filter were installed on the oxygen machine. The results show that the modification is effective and reliable, and the oxygen machine can pass the EFT test at 1 and 2 kV voltage level. Table 7.4 shows the contrast table before and after EFT Amendment.

7.3.2   Analysis of EFT Case 2

Product Electromagnetic Compatibility Modification Standard: China Standard: YY-0505—2012.

1. Product Introduction and the Description of Problem

As shown in Fig. 7.23, the vascular imager produced by a company is a display device that can display the thickness, shape and layout of veins in real time, to help medical personnel find veins. Achieve non-invasive, non-nuclear medical radiation (X-ray, γ-ray, etc.), reducing the suffering of specific patients and doctor–patient disputes.

The angiography apparatus is designed based on the principle that hemoglobin has strong infrared light absorbing ability. Oxygenated hemoglobin and dc-hemoglobin are more capable of absorbing infrared light than other tissues. Therefore, by sensing the intensity of reflected infrared light, and through a series of signal processing, the shape of the blood vessel can be displayed on the display. The instrument consists of an infrared light source generator, a filter system, a CCD sensor chip, an image signal processing module, and a display. The infrared source has a wavelength of 960–980 nm. Figure 7.24 shows the structure of the vascular imager.

"GB/T 17626.4—2018 electromagnetic compatibility test and electric fast pulse group anti-interference test" (equivalent to IEC 61000-4-4 standard) standards specify fast transient burst test standards and test methods, as shown in Table 7.5.

Due to the application of RF devices or switching devices and digital circuit control modules in the system, the vascular imager does not only generate a large amount of electromagnetic interference during normal operation, but also is affected by external electromagnetic environment noise. The Standard YY-0505—2012 uses the electric fast pulse group generator to carry out the experimental voltage peaks at positive and negative 1 kV and 2 kV, respectively, at the power supply port, the repetition frequency is 5 kHz, and the test duration is 60s. After applying an electrical fast burst to the power supply end of the device, the vascular imager failed the EFT immunity test. The specific phenomenon is that the display screen flashes during the

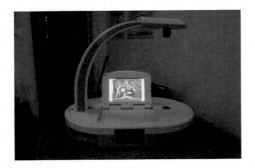

Fig. 7.23 Vascular imaging apparatus

Fig. 7.24 Vascular imager structure

Table 7.5 Burst test level

Level	Power port and protective earth (G)		I/O signal, data and control port	
	Peak voltage (kV)	Repeat frequency (kHz)	Peak voltage (kV)	Repeat frequency (kHz)
1	0.5	5/100	0.25	5/100
2	1	5/100	0.5	5/100
3	2	5/100	1	5/100
4	4	5/100	2	5/100
X	Undefined	Undefined	Undefined	Undefined

Note X is an open level, which must be specified in special equipment technology

Fig. 7.25 Vascular imaging device fault

test. And only after power off and restart can return to normal. Figure 7.25 shows the fault diagram when the vascular imager is subjected to the EFT interference signal.

2. EFT Inhibition Theory Analysis

EFT mainly interferes with the power port and transformer port of the vascular imager device in a conductive manner. The printed circuit board contains several sensitive electronic components such as MOSFETs and diodes. When these circuit modules are subjected to an electric fast pulse group with a short rise time and a high repetition rate, the charge accumulates in the nonlinear energy storage element. It affects the function of the local sensitive component and thus interferes with the internal circuit of the therapeutic device, which seriously affects the normal operation of the therapeutic device. According to the principle analysis, first check the ground wire. When the grounding is normal, the display panel has a display problem, so it is necessary to check the power supply line again. After disassembling the rear casing, it can be seen that the system design has certain defects. The distance between the high-frequency switching device and the sensitive amplifier component is too close, causing the amplifier to receive too much high-frequency crosstalk. At the same time, the wiring between the boards is messy and too long. Several circuits equivalent to radiated interference antennas. The power lines on each PCB board are simply grounded, and there is no effective shielding. There is a motor for generating three oxygen machines inside. Since the motor is a nonlinear device, many interference signals of different frequency bands are coupled, and these signals are not shielded and thus are received by the display module and the output module of the vascular imager.

3. Analysis of EFT Rectification Methods

When an electrical fast pulse group is applied to a vascular imager, the effect of the EFT on the imager can be observed using an oscilloscope and a voltage probe. In the experiment, 2 kV EFT interference is applied to the power port, and a spike interference voltage occurs.

$$f = \frac{1}{\pi t_r} = 64\,\text{MHz} \tag{7.7}$$

$$\lambda = \frac{U}{f} = \frac{3 \times 10^8}{f} \tag{7.8}$$

1. Power Cable and Communication Cable

The ferrite ring exhibits a high-resistance state at medium and high frequencies, which is equivalent to a series differential mode inductance on a circuit transmission line, and this equivalent inductance is low due to a quality factor. During operation, the ferrite ring causes the pulsed magnetic field generated by the high-frequency pulse current entering the power line to form an eddy current on the ferrite ring, and the high-frequency energy is consumed in the form of thermal energy, so that the high-frequency differential mode interference signal is well suppressed. Different ferrite rings have different impedance characteristics, and the circuit needs to be calculated so that the cutoff frequency of the ferrite ring just falls near the frequency of the interference signal. The ferrite ring is an absorbent filter. Multi-turn winding can change the impedance characteristics of the ferrite ring. In addition, it can improve the filtering performance of the ferrite ring in the entire frequency band.

The ferrite ring plays a very important role in dealing with electromagnetic compatibility problems. The ferrite ring cannot only filter the environment of the environment, but also prevent the internal interference of the system from entering the space electromagnetic environment. Furthermore, it is easy for installation, so they are widely used. Figure 7.27 shows how the ferrite ring is added. Table 7.6 shows the comparison of EFT filter ring test result.

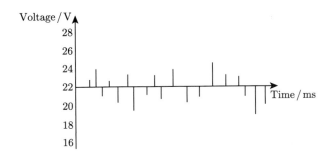

Fig. 7.26 EFT test result after modification

Fig. 7.27 Snap ferrite ring
at signal output

Fig. 7.28 Metal copper film
covering shielded signal line

2. Partial Shielding of Sensitive Circuits to Improve the Anti-Interference

There are many MOSFETs and electromagnetic relays on the main control board of the vascular imager. These devices generate a large amount of di/dt and dv/dt during the switching process. These periodic pulse signals can be coupled by electric field coupling, magnetic field coupling, conduction coupling. The common-impedance coupling method is coupled with the EFT interference signal into a nonlinear interference signal. The electromagnetic shielding is to avoid the occurrence of such coupling, thereby reducing the difficulty of eliminating noise. The principle of shielding is to seal the electromagnetic interference source with a metal material so that the external electromagnetic field strength is lower than the allowable value; or to close the electromagnetic sensitive circuit with a metal shielding material, so that the internal electromagnetic field strength is lower than the allowable value. In general, the shielding effectiveness is above 30 dB to be effectively shielded, and the calculation formula of shielding effectiveness is as follows:

Table 7.6 Comparison of 2kV EFT test results

Modification	Test results
Without ferrite ring	Not pass
Basic ferrite ring	Still many pulse harmonic interferences in the oscilloscope
Multi-turn wire-wound ferrite ring	Pulse harmonic interference is significantly less than single circle

Table 7.7 Comparison of 2kV EFT test results

Modification	Test result
Uncovered copper film	Not pass
Uncovered digital circuit chip	The device responds incorrectly and displays an error
Copper film covered digital chip	The treatment device is working properly

$$SE = 20 \lg \left(\frac{E_i}{E_t} \right) \tag{7.9}$$

E_i and E_t are the incident and transmitted electric field strengths, respectively.

In the circuit theory method, the incident field induces a current in the shield. These currents in turn generate additional fields, which cancel out the original incident field in some spatial regions. Such a processing method is called absorption loss, generally for low-frequency magnetic fields. In another way, some electromagnetic waves are reflected back from the surface, and this reflection loss is related to the impedance characteristics of the shield. The shielding method can effectively reduce the interference of the secondary radiation interference caused by the EFT pulse group through the switching device and the digital circuit to other parts of the circuit. Figure 7.28 presents the shielding method of the signal line using the metal copper film. Table 7.7 shows the EFT shield test result comparison.

4. Final Results After Modification

After the above analysis and demonstration, the ferrite ring, copper film, and EFT filter were added to the vascular imager, and the PCB-level grounding direction on the instrument was improved, from single-point grounding in series to single-point grounding in parallel. After the rectification is completed, all the measures are added and then tested. The results show that the modification measures are effective and reliable, and the vascular imager can work normally under the interference environment of 1 and 2 kV voltage level EFT test. Table 7.8 shows the comparison table before and after modification.

Table 7.8 Comparison table before and after modification

Voltage level	Before	After
1 kV	The sensitivity of the button is lowered and the indicator light is abnormally flashing	The button responds normally under the pulse group, the indicator light is normally illuminated, no flashing
2 kV	Communication often interrupts the sampling compressor without reaction, the screen appears abnormally blurred image, and downtime occurs	There is no blurry and unclear image on the screen, and the machine does not appear to be reversed or restarted

References

1. Zhai X, Wang J, Geng Y, et al. Transfer function and network synthesis of electrical fast transient/burst generator based on latent-roots method. IEEE Trans Electromagn Compat. 2010;52(4):790–6.
2. Cerri G, De Leo R, Primiani VM. Electrical fast-transient test: Conducted and radiated disturbance determination by a complete source modeling. IEEE Trans Electromagn Compat. 2001;43(1):37–44.
3. Wu J, Li B, Zhu W, et al. Investigations on the EFT immunity of microcontrollers with different architectures. Microelectron Reliab. 2017;76–77:708–13.
4. Zhai X, Geng Y, Wang J, et al. New method to model the equivalent circuit of the pulse generator in electrical fast transient/burst test. ICE Trans Electron. 2009;92(8):1052–7.
5. Carobbi CFM. Measurement error of the standard unidirectional impulse waveforms due to the limited bandwidth of the measuring system. IEEE Trans Electromagn Compat. 2013;55(4):692–8.
6. Wu Z, Qi D. The device parameters simulation of electrical fast transient generator. Energy Procedia. 2011;12:355–60.
7. Ling W, Yong LY, Li JZ. Assessment method for the testing system of electrical fast transient/burst immunity and its uncertainty. In: 2015 7th Asia-Pacific conference on environmental electromagnetics (CEEM), Hangzhou, 2015. p. 287–92.
8. Spangenberg R, Quenzer J. Electrical fast transient burst insertion loss measurement method. IEEE Lett Electromagn Compat Pract Appl. 2019;1(2):48–52.
9. Lv Z, Yang S, Li H, et al. EMC performance analysis and countermeasure of electric actuator control system-electrical fast transient/burst problem solving. In: 2019 4th international conference on mechanical, control and computer engineering (ICMCCE), Hohhot, China, 2019. p. 756–7563.
10. Guo Y, Chen X, Lin Y. Study on the immunity of electrical fast transient burst in rail transit. In: 2017 2nd IEEE international conference on intelligent transportation engineering (ICITE), Singapore, 2017. p. 304–8.

Chapter 8
Introduction of Other EMI Issues: CS, RS, and Surge

8.1 Conducted Interference Susceptibility (CS)

8.1.1 Generation Mechanism and Analysis of CS

EMI sources generate high-frequency voltages and currents. They cause charge and discharge of capacitance and stray capacitance of devices, such as converter valves, commutating reactors, transformers, and filters, thereby inducing pulse currents to form common mode interference noise. High-frequency current can induce high-order harmonic voltage on stray inductance and produce differential mode interference noise [1]. These noises propagate along the line and enter sensitive equipment through coupling or electrical connection to generate conducted EMI [2].

8.1.2 Implementation Standards and Test Methods of CS

The standard of CS test is IEC 61000-4-6:2006. This standard is about the requirement of conduction susceptibility of EMI signals, which is produced by electrical and electronic equipment to RE transmitters in the frequency range of 150 kHz to 80 MHz. The disturbed electrical and electronic equipment involved in this standard is coupled to the RF field generated by the RF transmitter through at least one connecting cable (such as power lines, signal lines, and ground lines). This standard specifies methods for coupling interference signals into the interior of a device under test using different coupling devices to simulate the EMI. The device under test may be exposed to the EMI.

The basic principle of RF field-sensing CS test is to couple the standard specified interference signal to the test cable using the corresponding coupling and decoupling device [3, 4]. The interference signal flows through the tested cable into the EUT and excites the corresponding interference field intensity inside the EUT (as shown in Fig. 8.1). Therefore, the measurement method is to simulate the electric and mag-

© Science Press 2021
Y. Zhao et al., *Electromagnetic Compatibility*,
https://doi.org/10.1007/978-981-16-6452-6_8

Fig. 8.1 Electromagnetic
field generated by the
common mode current on the
cable under test

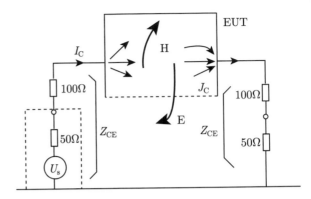

netic fields generated by the actual transmitter by the electric field and magnetic field formed by the device under the interference source. This interference can be represented by the near-field electric and magnetic fields formed by the voltage or current generated by the test device. Using coupling and decoupling devices, interference signals are applied to a cable under test and other cables are kept unaffected. Thus, the method can simulate the case where the interference source acts on different cables with different amplitudes and phases. In actual experiments, the desired signal level can be provided by a source with a nonzero impedance. The standard-specified common mode impedance is 150 Ω, which includes both the source impedance and the impedance of the cable connected to the EUT. The common mode impedance of the EUT and connecting cable must be tightly controlled to accurately predict the applied voltage. To achieve the desired common mode impedance and minimize the effects of other parametric changes on the test results, coupling and decoupling networks must be used to stabilize the impedance or decouple the auxiliary equipment. In order to reduce the impact of changes in parasitic coupling on the test results, the test arrangement should also be strictly controlled.

The cable to which the interference signal is injected may be a signal line, a control line, a ground line, or a power line. The current injection method is usually used to inject the specified interference signal. The common mode method is usually used to inject the interference signal, which is used to evaluate the susceptibility of the tested equipment to the common mode interference signal. The current injection method is required for the device under test, because in the actual RF electromagnetic environment, various types of EMI are easily generated. For unshielded conductors, useless signals are injected into conductors in common mode. For coaxial cables or shielded cables, useless signals are injected into outer conductors or shielding layers in common mode. In some cases, a portion of the common mode signal is converted to a differential mode signal, potentially masking the true common mode response.

The coupling and decoupling devices used in the tests are mainly coupling/decoupling networks (CDN), electromagnetic clamps, and current clamps. In order to study the anti-interference ability of electrical or electronic equipment to the conducted interference of the RF field induction, the basic method is to inject the useless

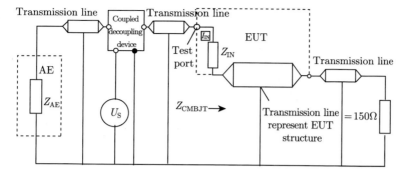

Fig. 8.2 Equivalent circuit of the injection method

signals into the wire and gradually increase the level until the EUT performance is reduced or the specified susceptibility level is reached.

From Fig. 8.1, the factors that have important influence on the current signal size of the injected equipment are coupling and decoupling device, characteristics of auxiliary equipment, test arrangement, etc.

In an actual test configuration, the coupling/decoupling device has a port that connects to the device under test and a port that connects to the auxiliary device. Each port is represented by a transmission line that represents the portion of the cable. Auxiliary equipment is represented by the common mode impedance. The device under test is represented by the input impedance of the test port, and the transmission line represents the structure of the device under test. The equivalent circuit of the actual test is shown in Fig. 8.2.

8.2 Radio Frequency Electromagnetic Field Susceptibility (RS)

8.2.1 Generation Mechanism and Analysis of RS

RF interference is EMI transmitted through space radiation [5]. The RF test interference is applied to the EUT by spatial radiation. The applied interference is a RF continuous wave signal of 80 MHz to 2 GHz. The interference field strength is 1–30 V/m. The modulation mode is 80% amplitude modulation of 1 kHz sine wave. The interference enters the internal circuit of the EUT, which may cause the input and output of the analog signal to deviate from the expected effect. It may result in control failure or disruption of digital circuit. This section provides an in-depth discussion of the formation mechanisms and hazards, implementation standards and test methods, and coping strategies.

Common RF radiation interferences include small handheld transceivers used by operations and security personnel, fixed wireless broadcasts, television transmitters, in-vehicle radio transmitters, radio communication equipment, radio monitoring equipment, and various industrial electromagnetic sources [6, 7]. This RF radiation floods the environment around people and affects most electronic devices in some way. In recent years, wireless telephones and other radio transceivers, such as Bluetooth devices and Wi-Fi devices, have become more and more widely used, and the frequency of these devices is between 0.8 and 3 GHz. This RF radiation also becomes a major component of radiation around us. The external radiated energy of the RF radiation is an essential element for the normal operation of the equipment, but it also becomes a source of EMI for other electrical and electronic equipment.

In addition to the intentional electromagnetic energy generated by the above examples, there are some devices that need to use RF signals during normal operation, such as switching power supplies, internal energy conversion and transmission through high-frequency switching signals, information technology equipment, and intelligent control devices. The internal intelligent control chip needs to coordinate all internal work through the clock signal. But the RF signal is inadvertently leaked out through the casing or various interface lines and becomes an EMI source of other electric and electronic devices through the space radiation.

These space radiations are directly accessed by the outer casing of the equipment under test (EUT) and received by their internal circuits. In addition, these space radiations are received by various interface lines of the EUT, which is transformed into conducted interference transmitted through these ports, and through these interface lines into the internal of the EUT. Due to the imbalance of the internal circuit of EUT, these common mode interference signals radiated from the space are converted into differential mode interference inside the EUT and superimposed on the input or output of the internal circuit. If the noise margin of the internal circuit is exceeded, it can affect the normal operation of the EUT and cause interference to the EUT.

Generally, the lower frequency radiation interference with longer wavelength is mainly received by the EUT's interface line into the device. The efficiency of directly entering the EUT through the device casing is very low. The higher frequency radiation interference enters the EUT through the EUT wiring port and can also be directly received by the EUT internal circuit through the EUT housing. Both ways of entering the device under test are effective. To produce a suitable spatial electric field strength, the lower the frequency, the greater the cost and volume of the measurement system. Considering the propagation mode of radiation interference and the cost of the measurement system, the susceptibility to RF radiation interference below 80 MHz is mainly measured by conduction injection through the port line of the EUT.

8.2.2 Implementation Standards and Test Methods of RS

1. Radiation Susceptibility (RS) Test System and Test Method Based on GTEM Room

To establish a stable electromagnetic field environment, there are two types of equipment recommended in the standard, the anechoic chamber, and the transverse electromagnetic wave chamber.

The anechoic chamber includes 3 and 10 m. It is a shielded room with absorbing electromagnetic wave materials on six sides. The electromagnetic field is generated by the double-cone antenna and the logarithmic period antenna. It is suitable for large-scale product testing, the cost is high, and the economy is not strong. The strip line and the transverse electromagnetic wave chamber are extensions of the coaxial transmission line. The volume is small, and the cost is low. The disadvantage is that the available test space is small, and it is only suitable for product testing of a very small size.

The giga transverse electric and magnetic field transmission (GTEM) cell is an extension device of the conventional transverse electromagnetic wave chamber in the frequency band. It is in principle a conical coaxial line structure with an air dielectric constant and a 50 W characteristic impedance at the feed point. The terminals at the end of the coaxial line are discrete resistors and absorbing materials to ensure broadband matching. After the cell inputs the RF signal, a transverse electromagnetic wave is excited around the core plate, and the field strength is proportional to the output voltage and the distance from the core plate to the ground. The GTEM cell provides very good field uniformity and field reproducibility in given workspace area.

The GTEM cell is also called gigahertz (GHz) transverse electromagnetic wave cell. It is a new type of electromagnetic compatibility testing equipment developed in recent ten years. The working frequency range can be from DC to several GHz or more, and the internal usable field area is large. In particular, the total price of the cell itself and supporting equipment is not too expensive. Most enterprises and institutions accept it. Therefore, GTEM cell has made a great progress in China, and it has become the preferred scheme for enterprises to test the susceptibility of RF radiation electromagnetic field for equipment with small size.

2. Principle of GTEM

The GTEM cell is a device designed according to the principle of coaxial and asymmetric rectangular transmission lines. In order to avoid reflection and resonance of internal electromagnetic waves, the GTEM cell is designed to have a tapered shape. The input end adopts an N-type coaxial joint, and then, the center conductor is flattened into a sector plate called a core plate. A rectangular uniform field region is formed between the core plate and the bottom plate of the chamber. In order to make the spherical wave (strictly speaking, the spherical wave is propagated from the N-type joint to the GTEM cell, but the spherical angle is approximated to the plane wave due to the small angle of the design), there are good transmission char-

Fig. 8.3 Shape of the
GTEM cell

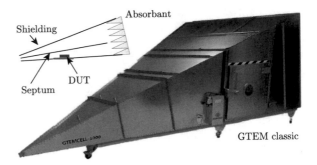

Fig. 8.4 GTEM voltage
standing wave ratio
characteristic diagram

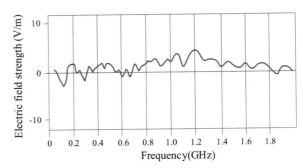

acteristics from the input end to the load end. The terminal of the core board is a
non-reflective terminal due to the use of a distributed resistance matching network.
The end face of the GTEM cell is also affixed with an absorbing material for further
absorption of electromagnetic waves at high frequency. Thus, a test area of uniform
field strength is created between the core and the bottom plate of the cell. During the
test, the measured object is placed in the test area. In order to prevent the uniformity
of the field from being affected by the placed object, the measured object should not
exceed 1/3 of the distance between the core plate and the bottom plate. Figures 8.3,
8.4, and 8.5 show the shape and typical operating characteristics of the GTEM cell.

The electric field strength in the GTEM cell is proportional to the input signal voltage U from the N-type connector and inversely proportional to the vertical distance
h from the core plate.

$$E = U/h \tag{8.1}$$

In a 50 Ω matched system, the relationship between the voltage of the core to the
backplane and the signal input power P of the N-type connector is:

$$U = \sqrt{RP} = \sqrt{50P} \tag{8.2}$$

The field strength is:

$$E = \sqrt{50P}/h \tag{8.3}$$

Fig. 8.5 GTEM electric
field intensity uniformity
curve

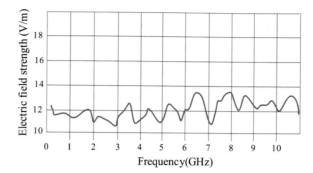

If considering the difference between the measured value and the theoretical value, the above formula should also be multiplied by a coefficient k, so the actual electric field strength is:

$$E = k\sqrt{50P}/h \tag{8.4}$$

If the same power is injected into the GTEM cell, the closer the position of the core plate is to the bottom plate, the larger field strength can be obtained. If the field strength in the same phase is generated, the input power required in the larger space is also larger.

For small objects measurement, it can be placed in the upper position of the GTEM cell, so that with a relatively smaller signal input power, a sufficiently high electric field strength can be obtained. It should be noted here that the height of the measured object cannot exceed 1/3 of the distance between the core board and the bottom plate at the selected position; otherwise, the field strength cannot be guaranteed to be stable and uniform.

3. Radiation Susceptibility Test System in GTEM Room

The RF electromagnetic field electromagnetic field susceptibility test in GTEM room is shown in Fig. 8.6. In addition to the GTEM outdoors, the main components of the RF electromagnetic field susceptibility test system include: RF signal generator, power amplifier, three-way connection, RF power meter, power probe and monitor, computer, etc.

In Fig. 8.7, when the RF signal generator sends a signal, it is amplified and injected into one end of the GTEM room (through the N-type coaxial connector, and the RF power meter is used to monitor the signal size to prevent excessive and too small signal levels produced), a strong uniform electromagnetic field can be formed between the core board and the bottom board. The electric field monitoring probe placed at the position of the object to be tested monitors the field strength, and then, the input signal level value is obtained through the computer. The signal source is directly adjusted to achieve the required field strength value.

The measurement and control software control the signal source to scan the frequency of the radiation field at a certain step size. There is also a video monitor (The

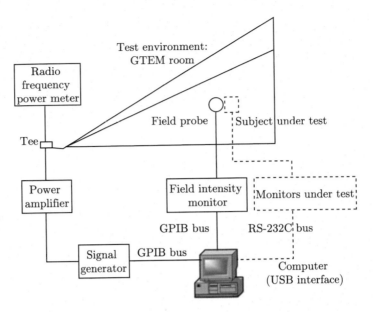

Fig. 8.6 Radiation susceptibility test system based on the GTEM cell

Fig. 8.7 Actually
constructed test system
layout

camera is installed in the GTEM cell, not shown in Fig. 8.3, and the tester passes the
video monitor in the GTEM cell.) to observe the operation of the object under RF
electromagnetic field interference.

The main steps of the test method are as follows:

(1) Closed-loop calibration, using the RF signal generator to generate the RF signal
of the required frequency range (typically 80 kHz to 1000 MHz, stepped by 1%
of the previous frequency value), after passing through the power amplifier feed
into the GTEM cell, comparing the current field intensity obtained by the field
intensity probe and the field intensity monitor with the required field intensity

(3 V/m, 10 V/m or other values). The output of the RF signal generator is automatically adjusted. According to the feedback closed-loop function, the field intensity in the small room is adjusted to the required value, and then, the regulated signal generator is output. The values are recorded automatically. When all frequencies are adjusted, the calibration work is completed. The output values of the signal generator corresponding to each frequency are obtained.

(2) Sweeping test, put the measured object into the same position of the GTEM indoor probe, and call the sweep program to make the GTEM room reproduce the field strength value of each frequency point reached. By the external monitor or the object monitor, monitors whether the measured object has abnormal operation or not. If there is an abnormality, the current frequency point is recorded in the sensitive frequency point list.

(3) Freezing test is a semi-automatic test. In this test, the list of sensitive frequency points is invoked, and the working condition of the tested object is carefully observed at each fixed frequency point to judge whether the tested object is qualified.

The use of the GTEM cell to construct a RF electromagnetic field radiation electromagnetic field susceptibility test system has the following outstanding advantages:

① The electric field produced by GTEM cell is much stronger than that produced by antenna. So, a relatively smaller RF power amplifier can generate a strong electric field, which greatly reduces the price of the entire test system. This is a very good RF radiation electromagnetic field susceptibility test scheme for devices that are not too large.

② Because the antenna is not needed in the test of RF radiation electromagnetic field with GTEM chamber, it can be conveniently used for automatic test, which greatly reduces the test time and the technical requirements for test operators.

4. Test System Software Introduction

As seen from Fig. 8.8, the test system software mainly consists of: signal generation, data reading, data analysis, data display, data saving, and report generation.

The function of the closed-loop calibration procedure is to maintain the field strength uniformity and value required for the test in the electromagnetic field within the GTEM cell. This calibration process is required before each test. All instruments and equipment are preheated to ensure that they are open and programmable. Instruments (Fig. 8.9) are selected, and programmable addresses are configured. Click on "Instrument Connectivity Test" to check whether the instruments in the system are connected properly (The light in the interface is connected through, when the light goes out, the connection fails). If the connection fails, check whether the instruments are in normal working state, whether the remote mode is open, and whether the instrument configuration is correct until the connection passes.

Then, place the field strength probe at the location where the object under test is going to be placed (it should also be on the "uniform domain" plane that has been calibrated). The parameter setting interface is shown in Fig. 8.10. According to the relevant standard requirements of the measured object, select the appropriate

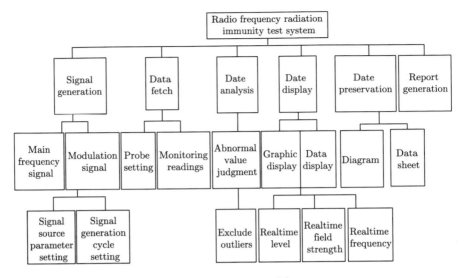

Fig. 8.8 Hierarchical relationship of system software modules

calibration field strength (1 V/m, 3 V/m, 10 V/m or custom) and the error allowed by the field strength calibration (for example 5%), select the starting and ending frequency of the test, the scanning step size, the frequency stepping mode (linear or logarithmic increment), the dwell time per frequency point, the parameters after calibration and the calibration data (each frequency point and corresponding signal generator RF level output) is stored in the database for later testing.

5. Test System Field Strength Uniformity Calibration

After all the hardware and equipment of the test system are built up, it is necessary to calibrate the field uniformity inside the GTEM room. The purpose of field strength uniformity calibration is to ensure that the field around the object under test is sufficiently uniform to ensure the validity of the test results. During the calibration process, no RF signal modulation is performed to ensure that the omnidirectional field strength probe indication is normal.

The concept of "uniform domain" refers to the vertical plane of an imaginary field, as shown in Fig. 8.11, where the field variation is satisfactorily small. When arranging the test, the surface to be irradiated of the object should be coincident with the vertical plane.

The magnitude of the field on the surface of 75% (at least 12 of the 16 points measured) within the specified area is within ±3 dB of the nominal value. The calibration procedure is as follows:

(1) Place the field strength probe at any point on the 16 points in the square (shown in Fig. 8.12).
(2) Feed the GTEM with RF power capable of transmitting a field strength of 3–10 V/m and simultaneously record two readings (power and field strength).

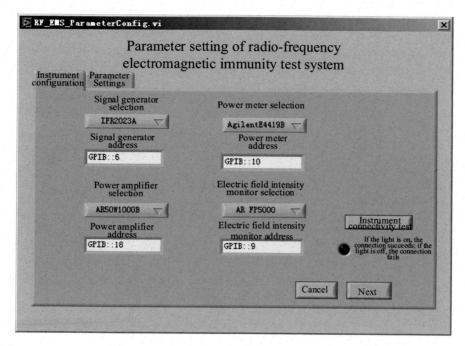

Fig. 8.9 Test system parameter setting diagram 1

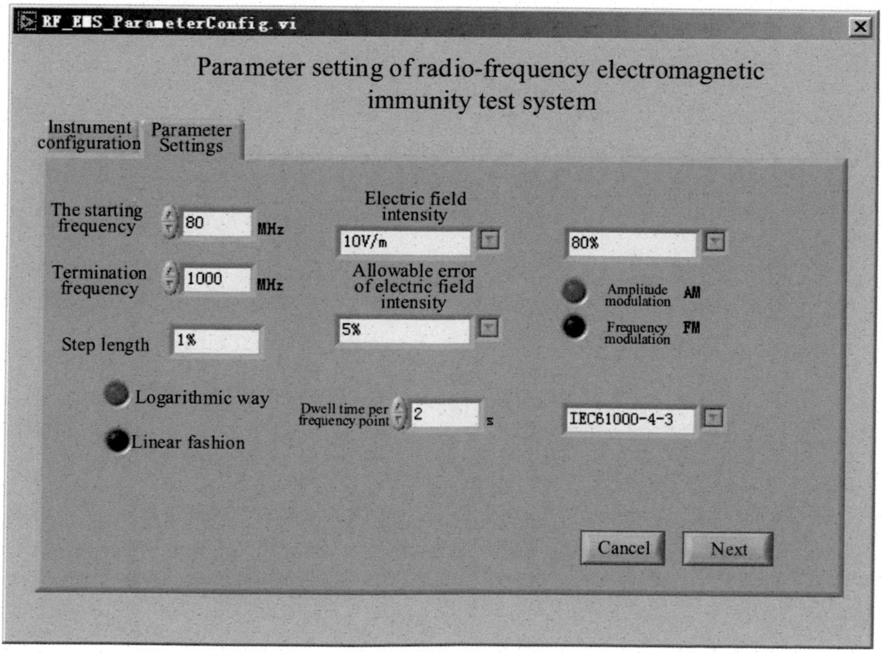

Fig. 8.10 Test system parameter setting diagram 2

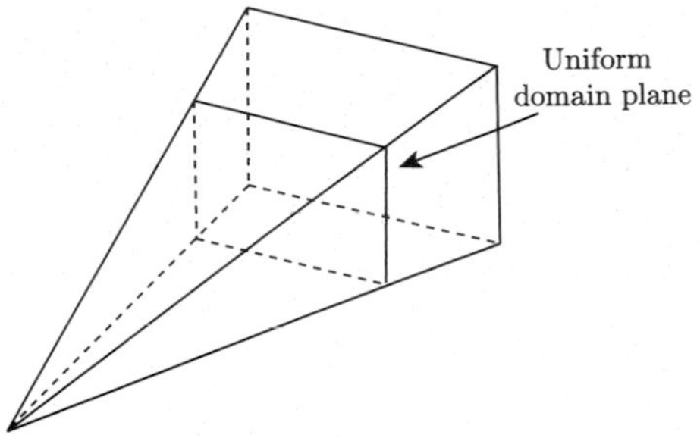

Fig. 8.11 Uniform field
plane of the GTEM cell

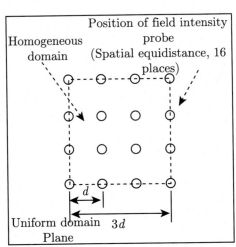

Fig. 8.12 Field strength
uniformity test method

(3) Measure and record the field strengths of the other 15 points with the same transmission power.
(4) Analyze the results of all 16 points, and eliminate 25% (total 4 points) of data with large deviation.
(5) The field strength of the reserved point should be within ±3 dB deviation.
(6) In the reserved point, the position of the lowest field strength is determined as a reference (ensure that the deviation is met in the range of 0 to +6 dB).
(7) From the relationship between input power and field strength, the power that must be transmitted for the required test field strength can be derived (for example, at a given point, the field strength generated by 1 mW power is 0.5 V/m, and the field strength generated by 36 mW power is 3 V/m).

This process is carried out after the equipment in the GTEM room is newly built or has undergone major changes (such as demolition and absorbing materials re-paste), or after the system has been operated for a long time, generally not required before each test.

After the field uniformity calibration is passed, the test system can be considered to be qualified and reliable, and can perform the RF electromagnetic field radiation susceptibility test specified in IEC 61000-4-3.

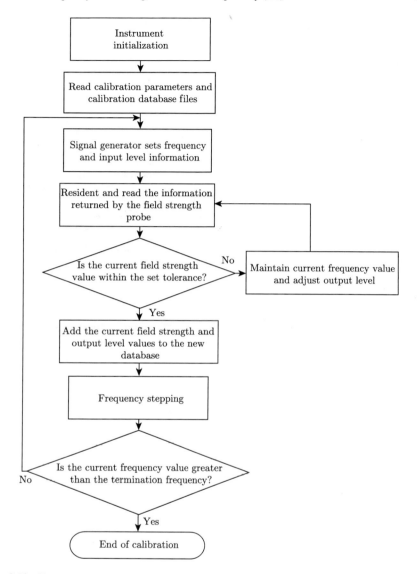

Fig. 8.13 Closed-loop calibration flowchart

Remove the field strength probe and place the object under test to ensure that it is in normal working condition. At this time, an external video surveillance system can be used to observe the operation of the EUT during the test. The closed-loop calibration flowchart is shown in Fig. 8.13, and a test flowchart is shown in Fig. 8.14.

The upper curve in Fig. 8.15 represents the RF level output at each frequency point, and the following curve represents the calibration field strength of each frequency point (within the allowable error of the selected calibration field strength). The right

Fig. 8.14 Flowchart of the
frequency sweep test

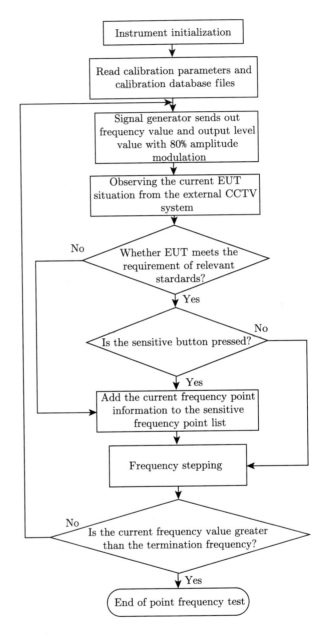

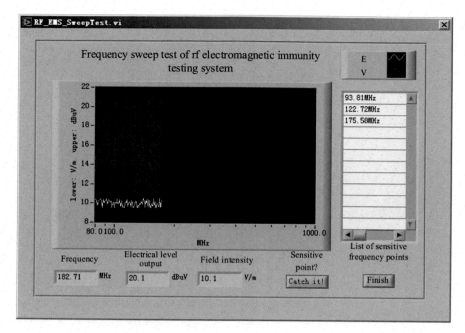

Fig. 8.15   Sweep test interface

side of Fig. 8.15 is a list of sensitive frequency points. The system automatically judges or when the EUT finds that the EUT is working abnormally at a certain frequency point. Click the "Catch it!" button below to display the current frequency value and RF output value is displayed in front of the button. (Provided by the calibration database). Add a list of sensitive points for use in the point frequency test.

6. Point Frequency Test Procedure

The point frequency test is the main vibration frequency point of the measured object (such as the frequency of the internal crystal oscillator of the measured object and the frequency multiplication frequency, which can be obtained by testing the radiation field strength of the measured object space), and the sensitive frequency points caught in sweep test are tested separately and more carefully. The dwell time is set manually to discover the real operation state of the Watt-hour meter under test at these frequency points. The point frequency test flowchart is shown in Fig. 8.16.

Fig. 8.16 Point frequency
test flowchart

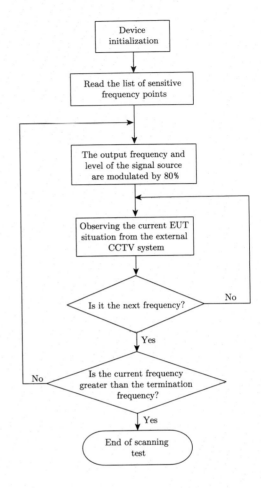

8.2.3 *Test System and Test Method of Radiation Susceptibility in Anechoic Chamber*

The interference from RF electromagnetic fields to equipment is often generated by equipment operation, maintenance, and safety inspectors using electromagnetic radiation sources such as mobile phones, radio stations, television transmitters, mobile radio transmitters (the above are intentional transmissions), the parasitic radiation (above is unintentional emission) produced by automobile ignition device, welding machine, thyristor rectifier, and fluorescent lamp which also produce RF radiation interference [8]. The purpose of the test is to establish a common standard to evaluate the ability of electrical and electronic products or systems to withstand RF radiated electromagnetic field interference. The test instruments include:

(1) Signal generator (main indicator is bandwidth, amplitude modulation function, automatic or manual scanning, retention time on the scanning point can be set, signal amplitude can be automatically controlled, etc.).

(2) Power amplifier (requires the field strength specified by the standard in the case of 1 m, 3 m, or 10 m. For small products, the 1 m can also be used for testing, but when the test results of the 1 m and the 3 m are different based on the results of 3 m).

(3) Antennas (double-cone and log-period antennas are used in different frequency bands, and composite antennas are used in the whole frequency band in foreign countries).

(4) Field strength test probe.

(5) Field intensity testing and recording equipment. Adding several basic instruments such as a power meter, a computer (including dedicated control software), and an automatic walking mechanism for the field strength probe can form a complete automatic test system.

(6) The anechoic chamber, in order to ensure the comparability and repeatability of the test results, the uniformity of the test site should be verified.

(7) Transverse electromagnetic wave chamber (TEM cell), strip line antenna, parallel plate antenna.

In accordance with the standard IEC 610004-3, the radiation electromagnetic field susceptibility test is to be amplitude modulated with a 1 kHz sine wave with a modulation depth of 80%, as shown in Fig. 8.17 (no modulation is required in earlier test standards). In the future, it is possible to add another keyed frequency modulation with a modulation frequency 200 Hz and a duty ratio of 1:1.

The test should be carried out in an anechoic chamber, as shown in Fig. 8.18. Monitor the operation of the sample with a monitor (or the signal from the sample can be used to indicate the working status of the sample to the measurement chamber, which is determined by a special instrument). There are antennas (including the antenna's lifting tower), turntables, samples, and monitors in the dark room. The staff, the instrument for measuring the crystallographic performance, the signal generator, the power meter, and the computer are placed in the measuring chamber, and the high-frequency power amplifier is placed in the power amplifier chamber. During the test, the wiring of the sample is very careful and should be recorded in order to reproduce the test results when necessary.

8.3   Lightning Surge

8.3.1   Generation Mechanism and Analysis of Lightning Surge

Lightning strike is a direct discharge phenomenon caused by the close contact between charged cloud layers or between a charged cloud layer and a certain part of

Fig. 8.17 Output voltage
waveform of the signal
generator

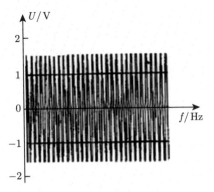

(a) Unmodulated RF signal $U_{pf} = 2.8V$, $U_{ms} = 10V$

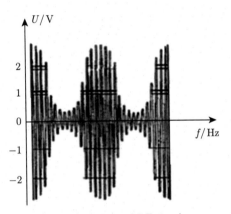

(b) Modulated RF signal

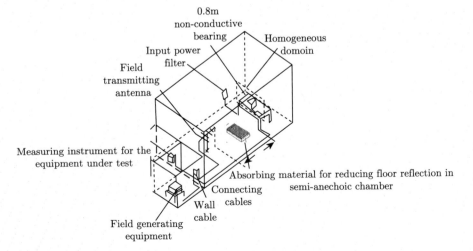

Fig. 8.18 RF radiation electromagnetic field susceptibility test configuration

Fig. 8.19 Lightning strike phenomenon

the ground. Figure 8.19 shows a strong lightning and loud sound and a large amount of energy transmission. The forms of lightning strikes are mainly direct lightning, conductive lightning, and induction lightning. With people's understanding of the formation of lightning bodies, people install lightning rods and lightning arresters on the top of buildings and lightning-sensitive areas to introduce electric charges close to the ground clouds into the ground for neutralization [9], thus effectively avoiding the occurrence of direct lightning strikes. The hazard of direct lightning strikes is greatly reduced. Lightning disasters are mainly caused by people and objects on the ground by lightning in the past and have developed into lightning waves transmitted by metal wires. In this section, the formation mechanism of lightning surge, the implementation of standards, and testing method. The corresponding strategies are discussed in depth.

Although people have already had better protection against the catastrophic damage of direct lightning and conducted lightning, indirect lightning (such as lightning strikes in the clouds, between clouds, or lightning strikes on nearby objects) can still induce surge voltage and current on outdoor overhead lines. In addition, in a power station or a switchyard, when a large switch is switched, a large surge voltage and current are induced on the power supply line. The common feature of these two surges is that the energy is particularly large (compared by energy, the electrostatic discharge is pico-joule, the fast pulse group is mill joules, and the lightning surge is several hundred joules, which is millions of times the energy of the first two kinds of interference), but the waveform is slower (microsecond level, while static and fast pulse groups are nanoseconds, even sub-nanoseconds), and the repetition rate is low. With the development of science and technology, semiconductor-integrated circuits and micro-control technology have penetrated almost all fields. Since semiconductor-integrated circuits cannot withstand overvoltage and overcurrent surges, surge impact damage accidents of devices using these components, such as intelligent household appliances, office automation products, industrial controllers, electronic computers, wired and wireless communication systems, have increased significantly. The main object of the lightning disaster has shifted from the disastrous damage to human beings and the environment to the damage to microelectronic devices and equipment.

Surge in the field of electromagnetic compatibility is generally derived from switching transients and lightning transients. The switching transient is related to switching interference of main power supply system, such as switching of capacitor banks and slight switching actions or compliance changes in the vicinity of the instrument in the distribution system. It is also related to resonant circuits associated with switching devices, such as thyristors and various system failures, such as short circuit and arc fault to the equipment group grounding system. Lightning strike transients are the main sources of surge (shock) voltage generated by lightning. Direct lightning strikes an external circuit (outdoor), and the injected large current flows through the grounding resistor or external circuit resistance impedance to generate a surge voltage. In buildings, indirect lightning strikes generate induced voltage and current on external conductor. The lightning current entering the ground directly to the ground discharge point nearby is coupled to the common grounding path of the grounding system of the equipment group. If there is a lightning protection device, when the protection device operates, the voltage and current may change rapidly, and coupled to the internal circuit, which still produces transient impact.

There are two ways of lightning striking electronic equipment: the first is that the high-energy lightning shock wave invades the equipment through outdoor transmission lines and connecting line-level power lines between devices, which damages the electronic equipment connected in series in the middle or terminal of the line. Secondly, lightning strikes the earth or grounding conductors, causing an increase of local instantaneous ground potential, thereby affecting nearby electronic equipment, which impacts the equipment and damages insulation to the ground.

Generally, the surge pulse has a longer rising time, a wider pulse width, and it does not contain a higher frequency component. It mainly enters the equipment through conduction. The effects of longitudinal (common mode) shock on the components of the balance circuit of the equipment include: damage to components that are connected between the ground and other insulating media; breakdown of components or other insulation between transformer turns, layers or lines to ground, which plays the role of impedance matching between circuits and equipment. Lateral (differential mode) shocks can also be transmitted in the circuit, damaging the internal circuit's capacitance, inductance, and semiconductor devices with poor impact resistance.

The extent of surge damage to components of equipment depends on the level of insulation and impact strength; insulation with self-restoring ability and ship is temporary. Once the shock disappears, insulation can be restored quickly. Some non-self-recovering insulating media cannot interrupt the operation of the equipment immediately if only a very small current flows after breakdown. However, with the passage of time, reverse breakdown occurs between the electrode and the emitter or between the emitter and the base, and permanent damage often occurs. For components vulnerable to energy damage, the degree of damage depends mainly on the current flowing through them and the duration.

8.3.2 *Implemental Standards and Test Methods of Lightening Surge*

Lightning strikes are a very common physical phenomenon. At least 100 lightning strikes per second occur in nature. In addition, switching operations in transmission lines can also generate many high-energy pulses that have a large impact on the reliability of electronic equipment. Therefore, many international and domestic standards have specified lightning surge tests. Surge immunity test for industrial process measurement and control devices GB/T 17626.5 (IEC 61000-4-5).

The anti-surge performance test of industrial process measurement and control devices is mainly to simulate the voltage and current surges caused by lightning strikes or switch switching in different environments and installation conditions. It establishes a common standard for assessing the power lines, input/output lines, and interference immunity of communication lines when subjected to high-energy pulse interference.

Lightning strike transients are mainly to simulate indirect lightning strikes, such as lightning strikes outdoor lines, a large number of lightning currents flow into external lines or grounding resistors to generate interference voltages, indirect lightning strikes (such as lightning strikes between clouds or clouds) induce on external or internal lines voltage or current, lightning strikes the adjacent object of the line, forming an electromagnetic field around it, causing a voltage to be induced on the external line, lightning strikes the nearby ground, and the ground current introduces interference through the common grounding system [10, 11].

Switching transients are analog: interference when the main power system is switched (such as a capacitor bank); when the same power grid, some switches near the device jump to form interference; switch thyristor devices with resonant lines; various system faults, such as equipment grounding short circuit or arcing between network or grounded systems.

The combined wave generator is shown in Fig. 8.20. U is a high- voltage source; R_C is a charging resistor; C_C is a storage capacitor; R_{S1} is a pulse duration forming resistor; R_{S2} is a pulse duration forming resistor; R_m is impedance matching resistor and L_r is rise time formed an inductance.

Fig. 8.20 Combined wave lightning surge generator circuit

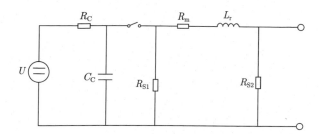

Fig. 8.21 Open circuit, short circuit waveform

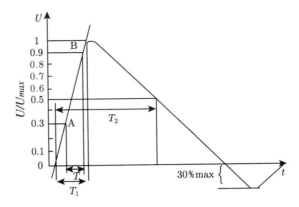

(a) Open circuit voltage waveform

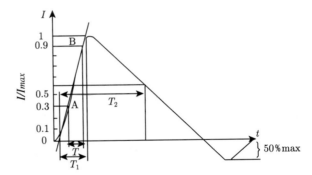

(b) Short circuit current waveform

The combined wave lightning surge electromagnetic compatibility test and measurement technology meets the GB/T 17626.5—2019 standard, and the surge (impact) susceptibility test meets IEC 61000-4-5:2014.

The open circuit voltage waveform and short circuit current waveform are shown in Fig. 8.21a and b, respectively. The 1.2/50 μs waveform parameters are specified in Table 8.1.

Combined wave power line capacitive coupling test method includes single-phase AC/DC power line capacitance coupling test (differential mode), single-phase AC/DC line capacitance coupling test (common mode), three-phase AC power line capacitance coupling test (differential mode), and three-phase AC power line capacitive coupling test (common mode).

The single-phase AC/DC power line capacitive coupling test (differential mode) steps are as follows: check whether the functional performance of the device under test is normal according to the technical requirements of the product; and then connect the device according to Fig. 8.22.

The test open circuit voltage and short circuit current waveform parameters are set according to the test category and test level specified by the product (EUT) technical

Table 8.1 1.2/50 μs waveform parameter specification

	Standard			
	GB/T 16927.1		IEC 469-1	
	Pre-wave time/μs	Half-peak time/μs	Rising time 10–90%/μs	Duration 30–50%/μs
Open circuit voltage	1.2	50	1	50
Short circuit current	8	20	6.4	16

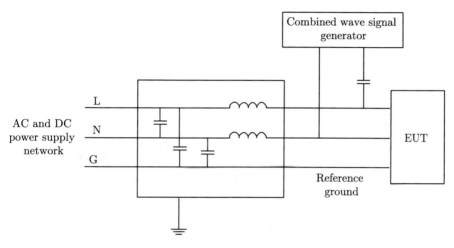

Fig. 8.22 Combination wave test (AC or DC power line-differential mode)

conditions. Then, the open circuit voltage wave combined wave signal generator is verified to open via the capacitor C output, 500 MHz oscilloscope, 100/1 attenuation probe connect the open end of the capacitor C. "Run" the combined wave generator, and the test generator output open circuit voltage indication error is within ±10%. Verify the short circuit current wave current coupling clamp on the combined wave signal generator output line and connect to the oscilloscope. The combined wave generator is short-circuited to the ground via the capacitor C. "Run" the combined wave generator, and the test generator output short circuit current indication error is within ±10%. Reconnect the combined wave signal generator, but does not run it. Disconnect test EUT and check that the residual-pulse voltage on the power supply line of the decoupling/coupling network should be lower than 15% of the peak value of the open circuit voltage waveform. Then, disconnect the power supply and check the residual power on the power supply line of the decoupling/coupling network input. The pulse voltage should be less than 15% of the peak value of the open circuit voltage waveform or not more than twice the peak value of the power supply voltage.

Then, reconnect the test equipment and the test product EUT. The combined wave signal generator does not work, and the test product EUT applies the nominal-rated

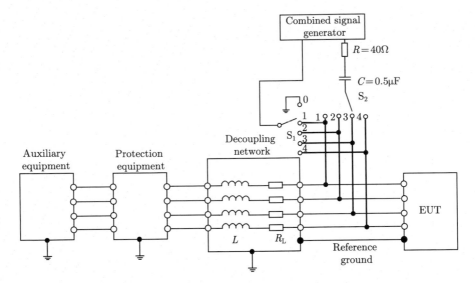

Fig. 8.23 Equipment connection of unshielded interconnect line capacitive coupling test

voltage. Check that the EUT function should be normal. Then, "Run" the combined wave signal generator to apply a surge pulse to the EUT (For AC power, the injection angle of the surge pulse applied to the supply voltage wave can be 0–360°; if there is no special regulation in the product standard, the test surge interference pulse can usually be synchronized at the zero crossing point and positive and negative peak point of the AC power waveform), test 1 time/min. Observe/check EUT function should be normal, and apply positive/negative interference pulses at each selected test site specified in the product standard, at least five times each test or according to technical conditions (the supply and demand sides determine). Then, disconnect all connections and re-examine the EUT function, performance should be normal. Then, judge the test phenomenon and results according to the test function status level specified in the product standard. Record the test results and prepare the submitted test report.

For the other three test methods, refer to the single-phase AC/DC line capacitive coupling test (differential mode).

Unshielded interconnect line capacitive coupling test methods include line-to-line (differential mode) tests and line-to-ground (common mode) tests. The line-to-line (differential mode) test shall be arranged and connected according to Fig. 8.22. The test method refers to the single-phase AC (or DC) power line capacitive coupling test (differential mode). S_1 and S_2 of Fig. 8.23 1, 2, 3, 4 and two combinations were tested separately. The line–ground (common mode) test shall also be arranged and connected according to Fig. 8.22. The test method refers to the single-phase AC (or DC) power line capacitive coupling test (differential mode). S_1 and S_2 of the Fig. 8.23 1,2, 3, 4 and two change Fig. 8.22 to Fig. 8.23 test the reference ground separately.

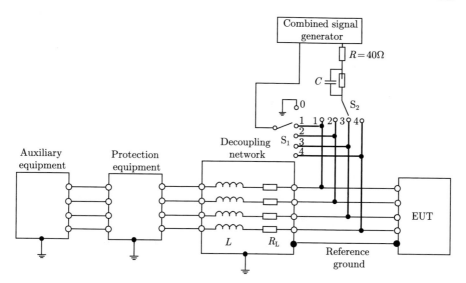

Fig. 8.24 Equipment connection of unshielded asymmetric working line gas discharge tube coupling test

    The unshielded asymmetric working line gas discharge tube coupling test method also includes line-to-line (differential mode) test and line-to-ground (common mode) test. Among them, the line-to-line (differential mode) test shall be arranged and connected according to Fig. 8.24. The test method refers to the line-to-line (differential mode) test of the unshielded interconnect line capacitive coupling test method, and in Fig. 8.24. 1, 2, 3, 4, the two-by-two combinations were tested separately. The line-to-ground (common mode) test shall also be arranged and connected according to Fig. 8.24. The test method shall be based on the unshielded interconnect line capacitive coupling test method line-to-line (differential mode) test, in Fig. 8.24 1, 2, 3, 4 of S_2 respectively, to test the ground.

    The unshielded symmetrical working line gas discharge tube coupling (communication line) test methods include line-to-line (differential mode) test and line-to-ground (common mode) test. The test shall be arranged and connected according to Fig. 8.25. Among them, in the line-to-line (differential mode) test, S_1 is set to 0 position for testing. The test method refers to the unshielded interconnect line capacitive coupling test method line-to-line (differential mode) test. The line-to-ground (common mode) test S_1 is set to 1, 2, 3, and 4 (ground), respectively. The test method refers to the unshielded interconnect capacitance coupling test method the line-to-line (differential mode) test.

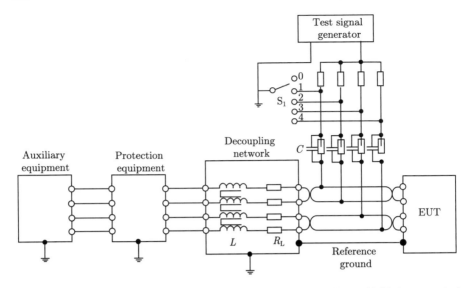

Fig. 8.25 Coupling of gas discharge tube coupling test equipment for unshielded symmetrical working (communication) line

References

1. Mi JC, Yang H, Yang PG. Proficiency testing in the field of electromagnetic compatibility. Saf EMC. 2018;1:27–30.
2. Hu JQ. Methods of conducted immunity test site calibration method. Electron Qual. 2015;7:71–3.
3. Guo Y, Gong J. Measurement methods and solutions of immunity to conducted disturbances. Digit Commun World. 2015;5:9–11.
4. Zang C, Su ZW, Li S, et al. Research on verification of EMC immunity test arrangement. China Med Devices. 2019;34(9):20–3.
5. Ye CQ, Wang J, Xie XC, et al. Summary of intermediate check on immunity test of electromagnetic compatibility. Electron Qual. 2019;12:77–83.
6. Xu P, Xu L, Chen Q, et al. The anti-interference solution and validation of EM radiation based on simulation. J Microwaves. 2014;30(s2):58–60.
7. Zheng WS, Wang W, Li HY, et al. Research on a new intermediate checking method for conducted disturbance and immunity test systems. Saf EMC. 2015;2:33–7.
8. Lu ZH, Liu PG, Liu JB. On-site measurement method for radiation emission based on adaptive anti-interference technology. Syst Eng Electron. 2012;34(2):243–8.
9. Kong QQ, Song QJ, Tan C, et al. Analysis of the new surge (shock) immunity test standard. Electron Qual. 2020;2:44–7.
10. Yang M, Chen JW, Xiong J. Study on the surge suppression technique. Insul Surge Arresters. 2005;6:28–30.
11. Tian HJ, Qian YX, Ma X. Study on calibration techniques of surge waveform in EMC testing. Saf EMC. 2012;3:75–8.

Printed in the United States
by Baker & Taylor Publisher Services